Behaviormetrics: Quantitative Approaches to Human Behavior

Volume 8

Series Editor

Akinori Okada, Professor Emeritus, Rikkyo University, Tokyo, Japan

This series covers in their entirety the elements of behaviormetrics, a term that encompasses all quantitative approaches of research to disclose and understand human behavior in the broadest sense. The term includes the concept, theory, model, algorithm, method, and application of quantitative approaches from theoretical or conceptual studies to empirical or practical application studies to comprehend human behavior. The Behaviormetrics series deals with a wide range of topics of data analysis and of developing new models, algorithms, and methods to analyze these data.

The characteristics featured in the series have four aspects. The first is the variety of the methods utilized in data analysis and a newly developed method that includes not only standard or general statistical methods or psychometric methods traditionally used in data analysis, but also includes cluster analysis, multidimensional scaling, machine learning, corresponding analysis, biplot, network analysis and graph theory, conjoint measurement, biclustering, visualization, and data and web mining. The second aspect is the variety of types of data including ranking, categorical, preference, functional, angle, contextual, nominal, multi-mode multi-way, contextual, continuous, discrete, high-dimensional, and sparse data. The third comprises the varied procedures by which the data are collected: by survey, experiment, sensor devices, and purchase records, and other means. The fourth aspect of the Behaviormetrics series is the diversity of fields from which the data are derived, including marketing and consumer behavior, sociology, psychology, education, archaeology, medicine, economics, political and policy science, cognitive science, public administration, pharmacy, engineering, urban planning, agriculture and forestry science, and brain science.

In essence, the purpose of this series is to describe the new horizons opening up in behaviormetrics — approaches to understanding and disclosing human behaviors both in the analyses of diverse data by a wide range of methods and in the development of new methods to analyze these data.

More information about this series at http://www.springer.com/series/16001

Shizuhiko Nishisato · Eric J. Beh ·
Rosaria Lombardo · Jose G. Clavel

Modern Quantification Theory

Joint Graphical Display, Biplots, and Alternatives

Springer

Shizuhiko Nishisato
Univesity of Toronto
Toronto, ON, Canada

Rosaria Lombardo
Department of Economics
University of Campania Luigi Vanvitelli
Capua, Caserta, Italy

Eric J. Beh
School of Mathematical
and Physical Science
University of Newcastle
Newcastle, NSW, Australia

Jose G. Clavel
Departament of Quantitative Methods
Universidad de Murcia
Murcia, Spain

ISSN 2524-4027 ISSN 2524-4035 (electronic)
Behaviormetrics: Quantitative Approaches to Human Behavior
ISBN 978-981-16-2469-8 ISBN 978-981-16-2470-4 (eBook)
https://doi.org/10.1007/978-981-16-2470-4

This Springer imprint is published by the registered company Springer Nature Singapore Pte Ltd.
The registered company address is: 152 Beach Road, #21-01/04 Gateway East, Singapore 189721, Singapore

Foreword

Quantification theory has a long history and its popularity has reached every corner of the world during the past 100 years. It has drawn us into the fascinating world of data analysis and as such has played a key role as an attractive research tool for diverse areas of scientific disciplines.

During the first half of the last century, the foundation of quantification theory was firmly established by such researchers as Pearson, Gleason, Ramensky, Richardson, Kuder, Hirschfeld, Fisher, Guttman, and Maung; in the latter half of the last century, it flourished into routine methodology through concerted efforts by groups of researchers led by Hayashi (Japan), Benzécri (France), de Leeuw (the Netherland), Nishisato (Canada) and the trio of Young, de Leeuw, and Takane. There were of course a countless number of other outstanding individual researchers as well.

I am greatly honoured to write the foreword for this book. As a long-time researcher at the Institute of Statistical Mathematics (ISM), I am privileged to have worked with the late Dr. Chikio Hayashi who is well known for his theory of quantification and many other researchers at ISM. During my career, I spent several months as a visiting scholar at the University of Toronto with Prof. Nishisato and in return, I hosted him as a foreign visiting professor at ISM. When I heard about his dual scaling, what came to my mind was: when we expand real space to complex space, thus enriching our field of mathematical exploration, dual scaling must have the same effect of expanding our scope through re-directing our attention from simple space to dual space.

I met Prof. José Garcia Clavel at a conference and learned that he too had spent his sabbatical year in Toronto with Prof. Nishisato. I finally met Prof. Eric J. Beh and Prof. Rosaria Lombardo at the 2017 IFCS conference in Tokyo, whose fame I had known through their highly acclaimed book on correspondence analysis from Wiley.

By then, I was supporting Nishisato for his battle over the joint graphical display and learned that the three co-authors of the current book were also behind him. Nishisato reminisces in Chap. 1 the memorable CGS scaling debate at the 1989

IFCS meeting in Charlottesville, Virginia. I feel proud to say that I was there, and it was the first time I met him.

The current book offers an interesting mixture of two groups of researchers, Nishisato-Clavel and Beh-Lombardo. I heard that Nishisato had met Clavel at a conference in Barcelona where he was writing his Ph.D. thesis under the supervision of Michael J. Greenacre. As my generation knows, Nishisato is one of the last persons from the old school of traditional psychometrics, while Clavel is a contemporary researcher, gifted with modern technology. These two researchers with very different backgrounds have been working in unison for the last 20 years or so.

Beh and Lombardo are statisticians by training who have been exceptionally productive over the past two decades. It is my belief that they will play the role of the leaders in the development of quantification theory to the next level of advancement.

The four authors from Canada, Australia, Italy, and Spain have combined their unique talents in writing this book. The book itself is a collection of essays and technical writings on quantification theory. As a seasoned researcher, Nishisato offers his personal reminiscence over the history of quantification theory, where we see rare personal observations of the past researchers and their work. He then directs his focus on the joint graphical display of quantification results, a topic that has been ignored for decades; see his meticulous effort to solve the problem. Clavel and Nishisato present cluster analysis (hierarchical, partitioning, bi-clustering) as an alternative to the joint graphical display, where their main task is to explore groupings of row and column variables in dual space. Then, Beh and Lombardo present biplots as yet another alternative to the joint graphical display and expand their expert writings of other important topics of quantification theory.

This book represents a unique collaboration of two groups of researchers with different backgrounds, diverse viewpoints, and superb presentations. Their collaboration is very successful and the book demonstrates their own unique talents. As a whole, this is a very informative and uniquely helpful book as a technical guide for modern quantification theory. I would strongly recommend this book to many researchers in diverse disciplines.

Tokyo, Japan
October 2020

Yasumasa Baba

Preface

This book is a product of the collaborations by researchers from four countries with different backgrounds. The first contact was made when Beh and Lombardo published a highly innovative book, entitled *Correspondence Analysis: Theory, Practice and New Strategy* (Wiley, 2014), and Nishisato reviewed it (*Psychometrika*, 2016). With this background, Beh, Lombardo, Clavel (Nishisato's collaborator for some 25 years), and Nishisato proposed a session and presented papers at the IFCS (International Federation of Classification Societies) meeting in Tokyo in 2017. By then, through a number of correspondence, our close friendship was firmly forged.

The idea of writing a book together emerged through unfortunate and fortunate events. Beh was a reviewer of Nishisato's paper on joint graphical display which was, according to Beh, "controversial" but warranted a broader discussion because the issue had been largely ignored for decades. He recommended the paper for publication since he did not see strictly speaking anything wrong with it, the paper was well written and well argued. In spite of Beh's strongly positive review, the paper was unconditionally rejected as "fundamentally wrong". In the meantime, Nishisato successfully tested his solution to the long plagued problem of joint graphical display in quantification theory, the topic of his paper, at a conference. What happened then was sheer luck; although the review process of the aforementioned journal was strictly double blind, Beh could tell the identity of the author from the writing style and contacted Nishisato with strong encouragement. From these unfortunate and fortunate events, an idea of writing a book emerged, and two pairs of collaborators (Lombardo and Beh; Clavel and Nishisato) finally reached a decision to put our different ideas together into a book. After the 2017 IFCS meeting in Tokyo, we received encouragement from Akinori Okada (Series editor), and we finalized our decision to publish a book as a joint work of the Nishisato-Clavel team and the Beh-Lombardo team.

So, this is a product of our forged friendship and the book is by no means a unified product, for Beh/Lombardo and Nishisato/Clavel represent two different schools of thought. Beh and Lombardo are frontier researchers in statistics and their work is highly technical, while Nishisato and Clavel are more practice-oriented. In spite of our different backgrounds, we have come together to highlight the pros and cons of different ways of thinking about the same problem.

We do not describe anything that is strictly new, but rather discuss various issues in essay and technical form from both sides of the fence. Due to the different flavours of the two partnerships, you will see distinct differences in how we have described the topics, while using the same notation.

Part I consists of six chapters. Chapters 1–5 are based on Nishisato's reminiscence on his endeavour over half a century of research career with a particular perennial problem of joint graphical display. Clavel and Nishisato will discuss cluster analysis as an alternative to joint graphical display in Chap. 6.

Part II consists of five chapters edited by Beh and Lombardo. Chapter 7 provides a brief outline of the inner workings of reciprocal averaging and its role in correspondence analysis, while some previously unseen variants of reciprocal averaging are proposed in Chap. 8. Chapter 9 provides a brief historical introduction to biplots. Further discussions on biplots are presented in Chap. 10, although its focus is based on ordered categorical variables and multi-way data quantification. Finally, Chap. 11 explores some new ideas to deal with over-dispersed categorical data and its visualization.

Quantification theory as known by many aliases will continue to evolve and will capture the hearts of many researchers. This is a book written with the collaboration of four international researchers with different backgrounds and viewpoints. We hope that you will find this book as a useful addition to your bookshelf.

Toronto, Canada
Newcastle, Australia
Capua, Italy
Murcia, Spain
November 2020

Shizuhiko Nishisato
Eric J. Beh
Rosaria Lombardo
Jose G. Clavel

Acknowledgements

First, we would like to thank Prof. Akinori Okada, the editor of the Springer Baheviormetrics series, for accepting our proposal for the current book and constantly encouraging us, and Mr. Yutaka Hirachi of Springer Japan for his guidance. Our special thanks go to Prof. Yasumasa Baba for his kind Foreword, which makes all of us feel uplifted to the first-class leading researchers! We now would like to take this opportunity to express our appreciation to those who always offered us their helping hands with understanding, watchful eyes, and affections.

Jose G. Clavel

There are many combinations of fantastic events that have made my contributions to this book possible. Some of them I know, but others have happened without my awareness. I would mention only three of them.

It is evident that I owe greatly to the co-directors of my Ph.D. degree: Dr. Joaquaín Aranda (Universidad de Murcia, Spain) and Dr. Michael J. Greenacre (Universitat Pompeu Fabra, Spain). Dr. Aranda, my mentor, was the Head of the Department of Quantitative Methods when I began my teaching career at Universidad de Murcia. In my view, he always had confidence in my possibilities and has offered me his clear guidance not only for my teaching but also for my life. The second mentor, Michael J. Greenacre, is a renaissance man of our times, who opened my mind to what a university professor should be and he generously shared not only his knowledge with me, but also his friends at Sant Fruitosós de Bages and researchers all over the world.

The second important event was, of course, a casual meeting with Nishi in the hall of Universitat Pompeu Fabra, Barcelona, in July 1993. We were then attending the European Meeting of the Psychometric Society. I cannot remember why, but I was alone for lunch and decided to invite him to a truly Spanish meal—he clearly looked like a Japanese professor being alone. We had lunch together, but I did not know who he was. We continued our lunch together during the conference. Only later a friend of mine told me that he was Nishisato of dual scaling. This accidental meeting was the beginning of our cooperative work and that simple Spanish luncheon was the event which made a decisive influence on my life; it brought me later to Toronto, and more specifically, to Saint Georges Road, and then Old Mill Road, where I found a home away from home; thanks Lorraine.

Those matters of enormous importance, however, are nothing in comparison with the magnitude of compassion involved in the next recipients of my heart-felt appreciation: my dearest parents who, through their love, joy, and generosity, have given me the best family in the world; Dodo, Manolo, Mariuca, Beatriz, Elena, Joaquín, Javier, and Ciuca, my siblings, with whom I share this unimaginably great luck and fortune.

Rosaria Lombardo

I would like to thank my sisters Giovanna and Savina and my brother Federico who, differently from me, are *artists* and often inspired me to do my work *creatively*, especially when visualizing data. I would also like to thank Eric for having involved me in this project, but especially for his genuine friendship, and Nishi and José for their tenacity, patience, and understanding over the last years.

Eric J. Beh

Over the past few years, I've had the pleasure of working with some wonderful people. The last few years have also been extremely difficult for me, with various health issues getting in the way. So I'd like to acknowledge the collegiality and friendship I've received from Pieter Kroonenberg who I deem to be a legitimate *quasi-Aussie*; thanks mate. Of course, I am extremely grateful for the patience and understanding I have received from Nishi and Jóse over the years and, especially, Rosaria, whose friendship I've always valued.

There are others that I would also like to acknowledge who have helped me greatly over a much longer period of time and probably haven't been told often enough. My wife, Rosey, and son, Alex, know how much they mean to me but I'd like to give a special shout out to my Mum (Donella), my sister (Emma), her wife (Suzi), and their son and my (favourite, and only) nephew Oli. I love you all and thank you for everything.

Shizuhiko Nishisato

Since my first publication in 1960, it is already 60 years. I am indebted to a countless number of people, in particular, my Japanese mentors Masanao Toda and Yoshio Sugiyama at Hokkaido University, my American mentors R. Darrell Bock and Lyle V. Jones at the University of North Carolina, and my Canadian colleague Ross E. Traub at the University of Toronto. I was lucky to have had international contacts from the early days, such as Gaul, Bock, Mucha, Ihm, and I. Böckenholt in Germany; Lebart, Morineau, Tennenhaus, Saporta, and Le Roux in France; Mirkin and Adamov in USSR; Gower and Goldstein in Britain; de Leeuw, Heiser, Meulman, van der Heijden, and Kroonenberg in the Netherlands; Lauro and Coppi in Italy; and too many names to list here in the USA, Japan, and Canada. I owe greatly to all members of the Psychometric Society since 1961. Born in Japan, I spent my student days in both Japan and the USA, and professional life in Canada. For my truly fulfilled life, I do not have sufficient words to thank my wife Lorraine (née Ford), son Ira, his wife Samantha Dugas, and dearest grandson Lincoln Dugas-Nishisato, as well as my sister Michiko Soma and brother Akihiko Nishisato in Japan. Finally, my sincere and heart-felt thanks to the three wonderful co-authors Pepe, Rosaria, and Eric, who helped me to fulfill my life-long wish.

The most important step for the publication of the book was provided by the editorial staff of Springer Singapore, and we would like to express our sincere appreciation to all the editorial staff for their meticulous editing and helpful advice. Through their devotion, we can now see this beautiful book.

Contents

Part I
Joint Graphical Display

Preface

Over the past century and a half, quantification theory (QT) has been presented in many papers and books in diverse languages. For the current stage of its developments, please refer to Beh & Lombardo (2014), which is an excellent compendium of our current knowledge—other reference books in English, French, and Japanese will be presented later.

In the quantification of a two-way table of data (*e.g.*, contingency tables), we use singular value decomposition of the data matrix, and as such Torgerson (1958) called our quantification procedure *principal component analysis of categorical data*. The traditional principal component analysis (PCA) (Pearson, 1901; Hotelling, 1933), however, is different from quantification theory in two aspects, namely (1) continuous data for PCA and categorical data for QT and (2) primarily uni-modal analysis for PCA and bi-modal analysis for QT. These differences have led to respective courses of development.

As for the traditional uni-modal analysis of PCA, a typical data set may be patients-by-medical measurements (*e.g.*, blood pressure, heart rate, body temperature), and the object of the analysis is to find multidimensional relations of these medical statistics, where patients are considered a random sample, hence no direct interest in analyzing individual patients. The main task lies in finding multidimensional coordinates of these medical measures. This is a straightforward mathematical problem and there is no theoretical problem in finding the Euclidean coordinates for these variables.

In contrast, the bi-modal analysis of QT deals with such a data set as collected from different age groups of people on their most preferred life styles out of say ten choices. In this case, the main object of analysis is to find multidimensional relations between two sets of variables (age groups and life styles), and we must find multidimensional coordinates for the two sets of variables. It is almost certain that age groups and life styles are correlated to some degrees. This correlation makes the graphical display of QT results much more complicated than that of PCA, for we must face at least two questions: (1) how many dimensions are needed to describe the complete relations between two sets of correlated variables, and (2) how we can find the Euclidean coordinates of two sets of correlated variables in common space.

Thus, one-mode and two-mode tasks lead to distinct tasks. Note that we have one set of variables for one-mode analysis and two sets of correlated variables for two-mode analysis. One set of multidimensional coordinates of medical statistics is the direct output of PCA, while we have two sets of multidimensional coordinates, which is the reason why we talk about *joint* graphical display for QT. It can easily be inferred that the two-mode QT output requires a space of larger dimensionality than the one-mode PCA output. How to expand the space for QT appropriately, however, is not a simple matter, and this is known as *the perennial problem of joint graphical display*.

In typical courses of introductory statistics, we learn a geometric interpretation of *correlation* between two variables: the two variables are expressed as two vectors and the correlation (Pearson, 1904) is defined as the cosine of the angle between the vectors. Then, assuming that the correlation between the rows and the columns is not 1, a single component of QT requires two-dimensional space. Our problem then is how to double the space for the QT outcomes. Keep in mind that we are interested in placing both rows and columns of the contingency table in *common space*.

In QT, we must find coordinates to accommodate two sets of correlated variables in expanded space. In the history of QT, however, this task of space expansion has been intentionally or non-intensionally avoided or even ignored for the sole reason of practical graphing. Unbelievable as it may sound, this avoidance of space expansion for QT has become the main-stream of QT and it has dominated the QT literature and history. The main aim of Part 1 is to revisit the currently popular compromised graphical procedure and replace it with a logically correct method for joint graphical display.

French plot is currently the most widely used graph. This name became popular in the 1970s and 1980s. It is also referred to as *symmetric scaling*. We can define symmetric scaling as a joint graph of principal coordinates of rows and principal coordinates of columns, assuming that the *row-column correlation is perfect*. The problem is that this assumption of perfect correlation does not generally exist nor is interesting at all. Thus, the method is one of the simplified and logically compromised graphical methods. It ignores the role of correlation between rows and columns in joint graphical display. Instead, it is a graphical method that overlays the graph for row variables over that for column variables. This is currently the most widely used practice and is logically not correct.

After a few attempts to rectify this time-honoured practice, most researchers have so far ignored the problem and have used it as a standard procedure and have moved away from the inherent problem. Thus, the problem of joint graphical display has been ignored for decades. For some researchers, it remained as a perennial problem.

In the current book, we will re-introduce Nishisato's 1980 work to show how to solve this perennial problem. But, why did the solution to this graphical problem remain unknown to most researchers for so many years? Sadly or tragically, there was a good reason for it as we will see it later.

Once we derive exact Euclidean coordinates for rows and columns in common space, we encounter another problem, that is, how to graph a configuration in multidimensional space. This is one major problem that must await future investigation and implementation. An immediate alternative is to rely on *dimensionless analysis*, of which one popular method is cluster analysis. Thus, we will consider cluster analysis as an alternative to joint graphical display. We will examine the merits and demerits of the two approaches, joint graphical display and cluster analysis.

In Part 1, Nishisato will present his view on the traditional quantification theory, starting with early days, then some background information for solving the perennial problem of joint graphical display, and a solution in terms of his theory of doubled common space in the first five chapters. Then, Clavel will join him to discuss cluster analysis as an alternative to graphical display in Chap. 6. This final chapter of Part 1 will contain the two authors' final words on Part 1.

Notes on Shizuhiko Nishisato

Born in 1935 in Sapporo, Japan. After BA and MA in experimental psychology from Hokkaido University, Japan, I went to the University of North Carolina at Chapel Hill (UNC) as a Fulbright student and obtained Ph.D. under a joint programme of psychometrics and mathematics. At UNC, I was taught by excellent professors (e.g., former Presidents of the Psychometric Society [Bock, Jones, Kaiser, Adkins Woods] and Hotelling), had great fellow students (e.g., Rapoport, Messick, Mukherjee, Wiesen, Smith, Das Gupta, Abbe (née Niehl), Gordon, Zyzanski, Cole (née Stooksberry), Kahn, and Norvick). I served as the only subject for Prof. T. G. Thurstone's project (the late L. L. Thurstone's wife) for an English proficiency test for foreign students, each session was always followed by tea and cookies with her at the Thurstone Psychometric Laboratory.

Since 1961, the Psychometric Society became my home ground for the next 60 years, for its annual meetings became my personal arena to forge a wide acquaintance with key researchers (e.g., Gulikesen, Horst, Guttman, Torgerson, Green, Coombs, Harman, Kruskal, J. D. Carroll, Luce, Jöreskog, Cliff, Bentler, Bloxom, Ramsay, Arabie, Hubert, Ackerman, Molenaar, Fischer, Young, de Leeuw, Takane, Heiser, Meulman, van der Linden, Cudeck, U. Böckenholt, Thiesen, and others).

After one year of post-doctoral work at McGill University, I was recruited by Ross Traub to the Ontario Institute for Studies in Education, the University of Toronto (OISE/UT) in 1967. The Department of Measurement and Evaluation at OISE/UT soon became one of the centres of psychometrics in North America.

I coined the name **Dual Scaling** in 1976, retired in 2000 as Professor Emeritus, the University of Toronto, and currently live in Toronto with my wife Lorraine. I Served as President of the Psychometric Society, Editor of Psychometrika, Fellow of the American Statistical Association, Fellow of the Japanese Classification Society, Year 2000 Distinguished Alumnus of the UNC Psychology Alumni Association, and President of the Metropolitan Toronto Japanese Family Services, and served 20+years on the editorial board of Springer Verlag for the German Classification Society Data Analysis Series.

Shizuhiko Nishisato
University of Toronto, Canada.

Notes on Jose G. Clavel

Like most of the economists of my time, I finished my degree without knowing a word about the analysis of categorical data. Of course, we were taught in the first week of Statistics I that there were attributes that could be summarized in pie charts and so on, but after that, my classes moved into the quantitative world for the subsequent topics of Econometrics.

Thus, like Vasco Núñez de Balboa, I also saw my Pacific Ocean, when I was asked to write a thesis on multidimensional scaling (MDS), from which I moved years later to Correspondence Analysis, Classification and Regression Trees, Dual Scaling, and so on. In those days, more and more friends started pouring into my office with all types of data (not only economics and business data, but also stress data of race horses, non-cognitive skills data, LaLiga, and lately, covid-19 medical data). My policy of open doors, despite its interference of my work, has kept my mind fresh. It is like that "stay hungry and stay foolish" recommended by Steve Jobs.

In addition to those job requirements and my own curiosity, I have had the fortune to be in this age of computers. Starting with Framework and its spreadsheets, I followed the rainbow through Lotus, Excel, S-plus, Eviews, MATLAB, SPSS, Stata, R, and finally (so far) Python. This voyage had an important stop in Goregaon at the *Indira Gandhi Institute of Development Research*, Mumbai, India, where, during my sabbatical year, under the supervision of Dr. Dilip Nachane, a wise econometrician and friend, I learned how to program and write my own codes.

Looking back, I am sure that all these facts have contributed to my teaching life: my first-year statistics students will hopefully benefit from my background, as I take advantage of their desire to know more. I hope that the readers of this book will see that the categorical data have much more information inside than a simple pie chart.

José Joaquín García Clavel
Universidad de Murcia, España.

Chapter 1
Personal Reflections

Over half a century of his research career, Nishisato has observed the historical developments of quantification theory. He himself was involved in the heated arguments over the problem of joint graphical display of quantified results, the problem which lasted until recently when he solved it. So, please do allow him to reflect on his personal involvements in the controversy over the problem together with his overview of the early history of quantification theory.

It was the summer of 1960, in Tokyo, Japan, when Prof. Masanao Toda introduced Nishisato to Dr. Chikio Hayashi, one of the early pioneers of quantification theory. In 1961, with Hayashi's 1950 paper in his briefcase, Nishisato arrived as a Fulbright student at Chapel Hill, North Carolina, USA, and started his graduate work at the Psychometric Laboratory of the University of North Carolina. His supervisor was Prof. R. Darrell Bock, another pioneer of quantification theory, who was then teaching optimal scaling. Only after a while, Nishisato realized and understood that *Hayashi's theory of quantification* (Hayashi 1950) was essentially the same as *Bock's optimal scaling* (Bock 1960).

Later he learned that there were many other aliases for quantification theory. Outstanding among them was *French Analyse des Correspondances* (correspondence analysis), for it was unique with a strong emphasis on joint graphical display. To interpret the outcome of quantification, the French group used graphical display extensively, resulting in what we now know as *French plot*, or more neutrally *symmetric scaling* or *correspondence plot*. As mentioned earlier, this graphical method has a serious problem.

It has become a routine tool for visual display of quantification results. Because of its wide use, many researchers today are not even aware that this routine tool has had a long history of fierce debates, pro's and con's. We should note that some toughest critics of French plot would even denounce the use of French plot, but they were a minority and were ignored completely.

S. Nishisato et al., *Modern Quantification Theory*,
Behaviormetrics: Quantitative Approaches to Human Behavior 8,
https://doi.org/10.1007/978-981-16-2470-4_1

What is the problem with the French plot? The answer is that it does not provide an exact configuration, but is only a practical approximation to the true configuration. More concretely, the French plot employs half the space required for a complete description of the data configuration. In other words, a four-dimensional configuration of data is depicted in a two-dimensional graph. This is so because the French plot represents rows and columns of the contingency table as if the row-column correlation is perfect, that is, 1. But, we should realize that if the correlation is 1, we do not need a graphical display.

Critical views against any popular method are typically sidelined or suppressed, often into oblivion. This is the reason why Nishisato has chosen his own personal reminiscence as a useful vehicle to overview a historical background surrounding the joint graphical problem—his 1980 description of the total information in the contingency table has been more or less ignored, and when he tried to revisit the relevant theory in the book, a well-respected journal rejected his criticism on the current joint graphical display as "fundamentally wrong" for no obvious reasons.

We will see later that all the necessary information for finding multidimensional coordinates for correlated sets of variables was thoroughly discussed in his 1980 book. To understand the nature of his struggles, let us first go back to the early days of quantification theory, and then look at his solution to the perennial problem of joint graphical display in Chap. 5.

1.1 Early Days

In early 1960s, joint graphical display in data analysis was promoted by French researchers and the graphical display was uplifted to the level that it became almost synonymous to French *Analyse des Correspondances* (Benzécri et al. 1973). Since ideas of graphical display go back almost to the beginning of quantification theory, let us look at its early history first.

The birth of quantification theory goes as far back as the early years of the twentieth century (see Nishisato 2007a). First, we see an international group of ecologists interested in optimal mapping of two sets of variables (e.g., plants and environments) using gradient methods (e.g., Gleason, Lenoble, Ramensky), then we see similar developments advanced by famous social and statistical scientists in the 1930s and the 1940s (e.g., Edgerton and Kolbe, Ellenberg, Fisher, Guttman, Hirschfeld, Horst, Johnson, Maung, Mosier, Richardson and Kuder, Whitakker, Wilks), and then further developments in the 1950s and 1960s (e.g., Hayashi, Benzécri, R.D. Bock, Baker, Bouroche, Carroll, Escofier, Lebart, Lord, McDonald, Slater). These researchers established solid foundations for further developments of quantification theory. After 1980s, the number of publications on quantification theory increased substantially.

At the 1976 Annual Meeting of the Psychometric Society, Jan de Leeuw of the Netherlands, identified the following four groups of researchers as distinct promoters of quantification theory:

- **Japanese Quantification Theory Group**: Starting with Hayashi's papers in 1950 and 1952 in English, a large number of papers and several books were published, mostly in Japanese. The Institute of Statistical Mathematics (ISM), in Tokyo, played an important role in disseminating Hayashi's theory of quantification with Hiroshi Akuto's classification paradigm. In addition to those ISM researchers, there were also other outstanding researchers at several universities and research institutes throughout the country.
- **French Correspondence Analysis Group**: They published many relevant papers and books in French in the 1970s and 1980s, the best known being the two volumes by Benzécri and his many collaborators. Another outstanding French contribution is the journal *Les Cahiers de l'Analyse des Données*, devoted mostly to correspondence analysis. In terms of the number of active researchers, including those outside the Benzécri school, France was outstanding out of the four groups listed here.
- **Dutch Homogeneity Analysis Group**: Starting with de Leeuw's doctoral thesis at the University of Leiden, the Netherlands, in 1973, many young Dutch students published their doctoral theses on quantification theory and related topics in English from DSWO Press. Their contributions to the field were outstanding, and the University of Leiden was one of the centres of quantification theory, together with the above two groups. The other institutions in the Netherlands have also produced many outstanding researchers in psychometrics and related disciplines.
- **Canadian Group of Optimal/Dual Scaling**: For the purpose of dissemination to a large number of researchers in North America and abroad, Nishisato published five reports on Bock's optimal scaling and its generalizations (Nishisato 1972, 1973, 1976, 1979; Nishisato and Leong 1975), which culminated into his 1980 book. Since 1970, Nishisato and his students at the Ontario Institute for Studies in Education of the University of Toronto started presenting their studies on optimal/dual scaling and its generalizations to other types of categorical data at international conferences. Those days, Toronto was one of the centres of psychometrics in North America, with Ross E. Traub (Princeton University) in classic and modern test theory, Roderick P. McDonald (University of Queensland, Australia) in factor analysis and covariance structure analysis, Shizuhiko Nishisato (University of North Carolina) in optimal (dual) scaling, multidimensional scaling and other psychometric methods, and Raghu P. Bhargava (Stanford University) in multivariate analysis of discrete and continuous variables. The solid graduate programme was unfortunately terminated around 2000, and no more students were trained since then.

Out of these groups, the most colourful and vibrant was the French group, the main promoter of joint graphical display as its flagship. This is important to note because they had laid a solid mathematical foundation for quantification theory, together with a practical method for summarizing the quantification results in graphs.

In 2017, Nishisato chaired the Awards Committee for the conference of the International Federations of Classification Societies (IFCS), in Tokyo, and observed a dramatic change in submitted papers such that the reviews of relevant studies went only as far back as 10 years, as opposed to typically 30 to 100 years some 50 years ago. In old days, it was very important to identify the first author who developed a particular procedure. What a surprise for an old-timer! A proper review of the relevant studies used to be essential, and the paper with a scant review of relevant studies was typically rejected! In this modern time when any information is instantly available through the internet, old studies perhaps do not matter as much as half a century ago.

Here strictly for old-timers' sake, let us list those researchers one could regularly see at conferences in 1960s–1980s, namely those involved in quantification theory and related areas. As you will see below, the old academia was quite vibrant, and those were the days when international travels were extremely difficult, due to cost and visa restrictions. Hopefully, some readers would recognize their predecessors or mentors on this list. The names below are in the alphabetical order and represent only a sample of active researchers

- *Japan:* Adachi, Akiyama, Akuto, Aoyama, Asano, Baba, Haga, C. Hayashi, F. Hayashi, Higuchi, Inukai, Ishizuka, Iwatsubo, Kamisasa, Katahira, Kobayashi, Kodake, Komazawa, Kyogoku, Maeda, Maruyama, Miyahara, Miyano, Mizuno, Morimoto, Murakami, Nakamura, Nojima, Ogawa, Ohsumi, Okamoto, Otsu, Saito, Sakamoto, Shiba, Sugiyama, Takakura, Takeuchi, Tanaka, Tarumi, Tsuchiya, Tsujitani, Yamada, Yanai, Yoshino, Yoshizawa.
- *France:* Benzécri, Besse, Bouroche, Caussinus, Cazes, Choulakian, d'Aubigny, Daudin, Deville, Diday, Escofier-Cordier, Escoufier, Fenélon, Fichet, Foucart, Jambu, Kazmierczak, Lebart, Leclerc, Lerman, Le Calve, Le Roux, Marcotorichino, Morineau, Nakache, Pagés, Rouanet, Roux, Saporta, Schektman, Tabard, Tenenhaus, Tomassone, Trecourt, Vasserot (Note: Tenenhaus, Rouanet and Le Roux visited Nishisato in Toronto).
- *The Netherlands:* de Leeuw, Heiser, Israëls, Kiers, Kroonenberg, Meulman, Sikkel, Stoop, ten Berg, ter Braak, van der Burg, van der Heijden, van Rijckevorsel, van Schuur. Imagine that these young researchers were already in the frontiers of research!
- *USA, Canada, Australia:* Abbey, Arabie, Arri, Austin, Baker, Bechtel, Bentler, Bloxom, R.D. Bock, W. Böckenholt, Bradley, Bradu, Brown, J.D. Carroll, Chang, Chase, Cliff, Clogg, Coons, Cronbach, Curtis, Dale, de Sarbo, Edgerton, Evans, Fienberg, Franke, Gabriel, Gauch, B.F. Green, P.E. Green, Guttman, Hartley (Hirschfeld), Helmes, Horst, Hotelling, Hubert, Jackson, Jones, Katti, Kessell, Kolbe, Kruskal, Lawrence, Leong, Lord, McDonald, McKeon, Moore, Nishisato, Noma, Norvik, Odoroff, Olkin, Orlóci, Peet, Perreault, Prentice, Ramsay, Rao, Schönemann, Sheu, Singer, Sokal, Spence, Takane, Torgerson, Torii, Tucker, Wang, Wentworth, Whittaker, Young.
- *Britain:* Burt, Cox, Critchley, Digby, Everitt, Goldstein, Gower, Hand, Healy, Hill, Kendall, Krzanowski, Slater, Stuart.
- *Germany:* Baier, H. H. Bock, I. Böckenholt, Decker, Gaul, Ihm, Mucha, Pfeifer, Schader.
- *Italy:* Bove, Coppi, D'Ambra, Decarli, Lauro.

- *Spain:* Cuadras, Greenacre, Satorra.
- *Communist Countries:* Adamov, Aivazian, Mirkin from USSR (the Soviet Union) and Zudravkov from Bulgaria (Note: Nishisato was invited to the USSR and Bulgaria. In return, Mirkin and Zudravkov were invited to Canada).

These names are listed here to give the readers a rough idea of what the research scene looked like in those early years, say 1960s–1980s: it was a dawn of international collaboration. In spite of rare international travels, we observed closer personal contacts among researchers than nowadays—we used to send postcards to other researchers, requesting their reprints, and we almost certainly received the reprints after a long wait. Those reprints were valuable sources of information then.

In early days, most quantification-related studies were published as research papers and some of the frequently cited early studies in English are Pearson (1901), Richardson and Kuder (1933), Hirschfeld (1935), Horst (1935, 1936), Edgerton and Kolbe (1936), Wilks (1938), Fisher (1940) , Guttman (1941, 1946), Maung (1941), Mosier (1946), Hayashi (1950, 1952), Johnson (1950), Williams (1952), Bock (1956, 1960), Lord (1958), Slater (1960), McKeon (1996), McDonald (1968), Gabriel (1971), Hill (1973, 1974), Nishisato (1975), Teil (1975), Young et al. (1976), Nishisato (1978), Young et al. (1978), Nishisato (1980), and Takane (1980). An important paper by Tenenhaus (1982) was also available then before its publication. Please note that many studies in French, Japanese, and other languages were also published during these early years.

1.2 Internationalization

In the 1970s and 1980s, French researchers at INRIA (Institut National de Recherche et en Automatique) and other universities organized a number of international conferences in France, and many of us in the English-speaking countries were for the first time exposed to the enormous amount of rich and informative French work on quantification theory. Those days, French researchers were in the frontier of quantification research, and it was typical to hear such a comment from French researchers as "it was already discussed in Benzécri's lectures" on papers presented by English-speakers. At one conference, Brigette Escofier presented a talk and Nishisato commented that "we investigated the same topic ten years ago in Toronto." Guess what! Guttman, who was seated next to Nishisato, said "that was an excellent comment," and shook his hands firmly. This episode would tell us how dominant the French group was then. In those early days, the French group and Hayashi's group used to organize French-Japanese symposiums in France and Japan, promoting their cooperations.

The language barrier between French and English was still strong and it handicapped particularly English-speaking researchers. In this regard, the contribution of Greenacre was outstanding, for his 1984 book in English, based on his French PhD thesis in 1978, supervised by Benzécri, became almost instantly the main reference for both French and English quantification theory. In passing, Nishisato's 1980 book in English had been donated to Greenacre in response to the request from a member of his former thesis committee in 1981, which might (just might) have played a minor role in Greenacre's unprecedented success in 1984.

Similarly, in 1977, the German Classification Society (GfKl, Gesellschaft für Klassification) was hosted by Wolfgang Gaul at the University of Karlsruhe and he invited researchers from abroad (Note: according to Gaul, Nishisato was the first Japanese-speaking researcher he met, and later in 1990, the two published a paper in the Journal of Marketing Research, which is now known to be the first German contribution to this American journal). GfKl has invited many of those involved in quantification theory to present papers at its annual meetings. Thus, GfKl has decisively contributed much to the internationalization of research community. We owe much to such key German researchers as H.H. Bock, Gaul, Ihm, Opitz, Schadder, Mucha, Ritter, Baier, Decker, and many others, who were instrumental to opening our eyes to those researchers involved in classification research because they also invited many IFCS (International Federation of Classification Societies) members from many countries.

Another French contribution to internationalization is due to Tenenhaus, who published his famous paper in *Psychometrika* with Young (Tenenhaus and Young 1985). The paper was intended to introduce French correspondence analysis to the English-speaking researchers of the readers of *Psychometrika*. It was Tenenhaus' outstanding contribution, for the paper was originally written by Tenenhaus alone. Upon the rejection of the first submission, Young joined him. With more problems, Nishisato was recruited for further help to complete the paper. Towards the end of the joint work, however, Young decided to drop Nishisato over a dispute on joint graphical display, and the paper was at last published by the two authors. Later on, Tenenhaus apologized to Nishisato for his ignorance of Young's one-sided decision to drop him. This was another example where Nishisato's view on joint graphical display was sidelined.

1.3 Books in French, Japanese and English

Thanks to the internationalization of quantification theory, the first remarkable output was realized in the form of quantification theory books written in English. In terms of reference books, France and Japan were ahead of those researchers in English-speaking countries. To understand the extent of the international dissemination of quantification theory, let us look at books published in those three languages:

1. **Books in French**: Although many French books remained unknown to most English-speaking researchers, French contributions were outstanding. Many first-class researchers contributed to the theoretical developments of *Analyse de Correspondances*. Major books were: Benzécri et al. (1973), Caillez and Pagés (1976), Lefevre (1976), Bouroche (1977), Lebart et al. (1977), Jambu and Lebeaux (1978), Lebart et al. (1979), Saporta (1979), Bouroche and Saporta (1980), Nakache

(1982) , Cibois (1983), de Lagarde (1983), Escofier and Pagés (1988), Jambu (1989), Tenenhaus (1994), and Moreau et al. (1999)—this last book has a chapter on dual scaling, a contribution by Nishisato.

2. **Books in Japanese**: Many Japanese books also remained unknown to most English-speaking researchers. Let us glance at their contributions to quantification theory. Major books were: Hayashi et al. (1970), Hayashi (1974, 1993), Komazawa (1978, 1982), Nishisato (1982, 2007b, 2010), Hayashi and Suzuki (1986), Iwatsubo (1987) , Akiyama (1993), Komazawa et al. (1998), and Adachi and Murakami (2011). Quantification theory is discussed also in chapters of such Japanese books as Takeuchi and Yanai (1972), Nishisato (1975), Yanai and Takane (1977), Saito (1980), Yanai and Takagi (1986), Ohsumi et al. (1994), Takane (1995), Hayashi (2001), and Adachi (2006).
3. **Books in English**: Our readers may be familiar with books in English, but let us start from the early days of quantification theory: Lingoes (1978), Whittaker (1978), Lingoes et al. (1978), Nishisato (1980, 1994, 2007a), Gauch (1982), Meulman (1982), de Leeuw (1984), Greenacre (1984, 1993), Lebart et al. (1984), Nishisato and Nishisato (1984a, 1994), van der Heijden (1987), van Rijckevorsel (1987), van der Burg (1988), Kiers (1989), Koster (1989), Gifi (1990), van Buuren (1990), Weller and Romney (1990), Benzécri (1992), van der Geer (2003), Greenacre and Blasius (1994, 2006), Gower and Hand (1996), Clausen (1998), Blasius and Greenacre (1998), Le Roux and Rouanet (2004), Verboon (1994), Murtagh (2005), and Beh and Lombardo (2014).

We saw a worldwide dissemination of quantification theory since the middle of 1970s, as we can see in Nishisato's 1986 bibliography of 78 pages. Interested readers are also referred to later bibliographies by Birks et al. (1996) and Beh (2004).

To see overviews of later developments, look at such excellent review papers as Takane et al. (1991), Nishisato (1996) and Michailidis and de Leeuw (1998).

1.4 Names for Quantification Theory

Because quantification theory has appealed to a large number of researchers in different disciplines and countries, it has acquired many aliases such as gradient method, reciprocal averaging, simultaneous linear regression, Gutman scaling, Hayashi's theory of quantification, optimal scaling, principal component analysis of categorical data, correspondence analysis, homogeneity analysis, dual scaling and nonlinear multidimensional descriptive analysis. The list can go on as Nishisato (2007a), for example, lists 52 names.

These names came from the fact that researchers used different mathematical functions to quantify categorical data, to name a few: *maximize* (1) the row-column correlation of the contingency table, (2) the correlation ratio (the ratio of the between-sum of squares to the total sum of square), (3) the variance of the between rows and the between columns, (4) the regression coefficient of the rows onto the columns or vice versa, (5) the reliability coefficient alpha, or *minimize* (6) the discrepancy between row space and column space in the orthogonal coordinate system for bi-modal categorical data.

The optimization outcomes of these different criteria have, however, converged into solving singular value decomposition of essentially the same equation, namely singular value decomposition of the variance-covariance matrix of categorical variables. In essence, therefore, the procedure of quantification theory is closely related to principal component analysis (Pearson 1901; Hotelling 1933), the main difference being (a) whether the input data are continuous or categorical and (b) whether the analysis is uni-modal or bi-modal. For those interested in singular value decomposition, one can trace it back to Beltrami (1973), Jordan (1874), Schmidt (1907), Eckart and Young (1936), Johnson (1963), and Schönemann et al. (1965). Thus, let us accept the definition that quantification theory is singular value decomposition of categorical data in the bi-modal framework.

Notes on Dual Scaling:
Since the name "dual scaling" is closely related to the later discussion on joint graphical display, let us provide some background behind the name. In 1976, Forrest W. Young (USA) organized a symposium on "Optimal Scaling" at the annual meeting of the Psychometric Society, held at the Bell Laboratories, Murray Hill, New Jersey, where the speakers were Jan de Leeuw (the Netherlands), Gilbert Saporta (France), and Shizuhiko Nishisato (Canada) with the discussant Joseph P. Kruskal (USA), in order to exchange different views on quantification theory. At the discussion time, Joseph L. Zinnes (USA), a respected mathematical psychologist, raised the question whether the name optimal scaling was appropriate, because, he argued, any scaling method based on an optimization procedure was "optimal". After long debates, Nishisato proposed "dual scaling" for the reason that quantification theory is based on symmetric decomposition of data matrices. Since his proposal received general consensus of the large audience, this duty-conscious researcher adopted the name "dual scaling" in his 1980 book.

As of today, many researchers appear to prefer the name correspondence analysis to dual scaling. However, in the current book, we will see that dual scaling may be a more appropriate name than correspondence analysis for two reasons: (1) the optimal weights for rows and columns are symmetrically scaled (note: symmetrical = dual) and (2) the object of quantification is to find multidimensional coordinates for rows and columns in *dual space* (to be defined later). We will find it later that the solution to the perennial problem of joint graphical display can be found in this dual space, not in space for the contingency table, as used in correspondence analysis. These technical matters and terms will be explained later.

1.5 Two Books with Different Orientations

We should not forget that quantification theory was already very popular and well understood among Japanese-speaking and French-speaking researchers in those early days, thanks to both the Hayashi school and the Benzécri school.

Out of early books in English, there were two books with clearly distinct orientations, namely Nishisato (1980) and Greenacre (1984): Nishisato's book laid the algebraic foundations of quantification theory, while Greenacre's mathematically lucid book promoted the French tradition of graphical display.

Because Nishisato's book placed no emphasis on graphical display, some researchers did not even include his book in the literature review. In the midst of Greenacre's great popularity, there was one positive news to Nishisato: In the middle of 1980s, there was a search for an English book for Russian translation, and Nishisato's book was chosen by Mirkin, Adamov, and other committee members. The Russian translation was completed in 1990, when he was invited to Moscow. The contract for the Russian publication was signed between the University of Toronto Press and Finansi Statistika in Moscow. Shortly after that, however, an unexpected tragedy occurred: that historic collapse of the Soviet Union! Tragically and disappointingly the Russian translation was never published since the Moscow publisher was instantly closed down. Several subsequent attempts to raise funds also failed.

In retrospect, Nishisato should have extended his algebraic treatment of quantification theory to graphical display. He could then have explained why doubled multidimensional space was needed for joint graphical display, a point stated in his 1980 book, but not illustrated.

In contrast, the popular French plot offers a unidimensional graph for one-component data, and a k-dimensional graph for k-component data. In other words, the French plot depicts rows and columns in the same space. This simplified practice is not necessarily an oversight by French researchers, for we can later show that it is a clever way of reducing the space from the practical point of view, and that it often offers a good approximation to the exact Euclidean graph in doubled dimensions.

Relevant to this clever approximation is the statement by Lebart, Morineau, and Tabard (1977, see also Lebart et al. 1984) that one cannot not expect to obtain an accurate between-set distance (i.e., distance between a row and a column) from the joint graphical display by symmetric scaling (French plot). This is a very appropriate warning about French plot, and clearly shows that French researchers were well aware that the French plot is only an approximation to the true configuration. A sad aspect of the matter, however, is that their sensible warning or precaution has almost always been ignored by researchers in other countries. The fact remains, however, that symmetric scaling was promoted into routine use by many leading researchers, and that in the course of its popularization, it has become a standard method for graphical display. Keep in mind that it is not the exact method, but a compromise from the practical consideration.

Another possible reason why his 1980 book did not appeal to many researchers is because he proposed the name "dual scaling" for quantification theory. Strange as it may sound, a number of researchers avoided the name whenever quantification theory was referred to, or whenever correspondence analysis was discussed. Was it because such names as correspondence analysis, homogeneity analysis, and optimal scaling were more appropriate than dual scaling? Probably not. It is almost definitely because his book did not emphasize the joint graphical display.

As the readers will see later, the name dual scaling is perfectly justifiable and more appropriate than the other names: we will establish that "dual space" accommodates both rows and columns in common space, contrary to "contingency space" used by French plot which is not spacious enough to accommodate both rows and columns together.

1.6 Joint Graphical Display

The graphical display of quantification outcomes was an inviting topic for researchers who looked for simple and intuitively understandable visual illustrations of complex results. Keep in mind that we are primarily interested in the relations between rows and columns of the contingency table. The core of the problem then lies in the graphical display of rows and columns which are correlated. To derive coordinates for two sets of correlated variables, however, was not a simple task, and the search for the exact method has been short-changed or compromised, leading to the current situation where the logical framework is replaced with practical convenience. Part I is devoted to deriving a logical framework for solving the perennial problem of joint graphical display.

If we look at the history of quantification theory, it is curious why this problem has not been fully attended to. In early days of the history, French researchers, headed by Benzécri, offered a practical compromise, not the solution, to the perennial problem. An amazing fact is that this compromised graphical method has become an indispensable tool for quantification theory, leading to its status as the standard method for joint graphical display. This has been so since 1960, namely the past sixty years!!

As noted above, symmetric scaling is based on the assumption that the row and the column variables span the same space, that is, the rows and the columns are perfectly correlated. However, this does not normally happen in practice, and if they are perfectly correlated, we do not need graphical display of quantification results.

Because of the wide use of symmetric scaling, typical researchers are not even aware of this radical handling of the joint graphs, and symmetric scaling is considered totally legitimate. Part I of this book starts with the revelation of this amazing history and ends with a logical exposition on how one can derive Euclidean coordinates for both row variates and column variates in common space. This exposition, therefore, offers a giant step forward in the history of quantification theory.

The key to solve the perennial problem of joint graphical display was clearly discussed in Nishisato (1980). Unfortunately, the book was overlooked by Carroll et al. (1986, 1987, 1989), who proposed the CGS scaling as a solution to the problem of joint graphical display. The book was also ignored by Greenacre (1989), who denounced the CGS scaling from his own rationale. The starting point of the CGS scaling was fine, but the proponents never considered doubling the space. For this reason, the CGS scaling was nothing but French plot: the key to solve the problem was to double multidimensional space as Nishisato (1980) clearly demonstrated. Historically, the criticism of the CGS scaling by Greenacre (1989) led to the downfall of the CGS scaling.

It is already more than 30 years since then, but it is clearly shown in the current book that neither the CGS proponents nor opponent Greenacre saw what was at stake: the both parties argued over totally irrelevant matters. Once we revisit Nishisato (1980), it will become clear what is needed to solve the perennial problem of joint graphical display. The key one can find there is: *If we want to plot rows and columns in the common space, we must double the dimensionality of space, unless the correlation between rows and columns is perfect.*

Part I is fully devoted to explore this theory of doubled space. In the exposition, we will also demonstrate that the graphical method of symmetric scaling (French plot) is logically wrong, but it is an excellent and masterful compromise for exact joint graphical display.

We will first look at today's practice of joint graphical display, to learn that those currently popular graphical methods (i.e., symmetric scaling method and non-symmetric scaling method) are neither satisfactory nor logically correct. Then, we will derive exact principal coordinates for rows and columns of a two-way table in common space.

1.7 A Promise to J. Douglas Carroll

J. Douglas Carroll and Paul E. Green, two outstanding and pioneering researchers at the time, were joined by Schaffer to propose a method of joint graphical display (Carroll et al. 1986, 1987). After Greenacre's 1989 denouncement of the CGS scaling, the proponents tried to respond to the opponent, but without success (Carroll et al. 1989).

The proponents' defense of the CGS scaling and the opponent's criticism were totally out of focus, being unaware that the space must be doubled, the key described by Nishisato (1980) six years before the CGS scaling. An interesting point is that their arguments were centred only on half the necessary space where the traditional French plot and the CGS scaling provide identical results. Their arguments in retrospect, therefore, were totally non-productive. If both parties had considered doubled multidimensional space, the rebuttal over the CGS scaling had never been brought up. The only necessary step for the CGS scaling proponents to do was to expand the space.

In 1989, when Greenacre's critical paper was published, he was invited to give a talk on the paper at the IFCS (International Federation of Classification Societies) meeting, held at the University of Virginia, Charlottesville, organized by Hamparsum Bozdogan. At this conference, Greenacre denounced the CGS scaling again, making it so clear that it was Greenacre's triumph and the CGS's complete defeat. Immediately after Greenacre's talk with big applause, Carroll stepped outside the auditorium and Nishisato followed him. On the beautiful lawn of the university campus, Carroll was totally devastated by the humiliation in front of the large audience—Carroll was then one of the top researchers in this field and highly respected. Since Nishisato knew that Greenacre's criticism was out of focus, he promised Carroll that he would write a paper to defend it, which he did shortly after the conference.

His manuscript to defend the CGS scaling was sent to *Journal of Marketing Research*, where the CGS scaling papers and Greenacre's critical paper were published. The editor of the journal, however, returned the paper to the author without a formal review, stating that the CGS scaling had exhaustively been discussed by those leading scholars. It was obvious that the editor had not read the paper. Nishisato's inquiry about the non-review was totally ignored and left unanswered. This was an important lesson to Nishisato who assumed the editorship of *Psychometrika* a few years later.

Carroll, Green, and Greenacre were all frontier researchers and highly respected then and their arguments started well, but not extensive enough. It was also the time when the French plot was already accepted as a standard method of graphical display. With this background of time, the editor's decision of no review of the paper was truly regrettable, for the paper could have advanced the graphical approach one step further, without any shame for the three proponents of the CGS scaling. This was more than 30 years ago.

At this memorable Virginia conference, Nishisato met the next generation of outstanding Japanese researchers, among others, Baba, Okada, Imaizumi, and Mizuta. But, more importantly, he made a promise to a very despondent J. Douglas Carroll that he would defend the CGS scaling by writing a paper.

1.8 From Dismay to Encouragement

After 30 years since then, he remembered his unfulfiled promise to Carroll, and decided to write a paper to defend the motive of the CGS scaling and then to augment it into a solution to the perennial problem of joint graphical display. This was a simple task for him, for he knew exactly what to write. When he submitted his defending paper to a journal (not the Journal of Marketing Research), the editor instructed him to shorten the paper into a note and then into a short note. It was not a big deal to shorten it and re-shorten it, he thought, and obediently followed the editor's instructions. When a revised short note was submitted to the journal, the paper was rejected unconditionally by the editor, stating that it was "fundamentally flawed", as mentioned briefly in the Preface. This incredible rejection haunted the author, for he knew the journal well through his several publications under different editors.

Earlier in 2016, Nishisato presented a paper on the same topic at the annual meeting of the Japanese Behaviormetric Society, where he defended the same paper at a panel discussion with the chair Yasumasa Baba. He also presented a special talk on this topic at the annual meeting of the Japanese Classification Society and the talk was published in the society's journal (Nishisato 2019a). So, these events paved a way to fulfil his promise fully to the late J. Douglas Carroll.

Thanks to the support of the series editor Akinori Okada, the promise to the late J. Douglas Carroll will fully be fulfiled in Chap. 5 of Part I. Special thanks are due to Okada, then Yasumasa Baba for organizing a session at a conference (Nishisato 2016), where his promise to J. Douglas Carroll was successfully defended, and Ryozo Yoshino of the Japanese Classification Society who arranged for the publication (Nishisato 2019a). Chapter 5 is based on these papers (Nishisato 2016, 2019a, b).

Part 1 will start with a simple introduction to Euclidean graphs through principal component analysis, then move to principal component analysis of categorical data, with the discussion on differences between uni-modal and bi-modal analyses. This will take us to the key framework for introducing doubled multidimensional space, using simple numerical examples, and we will finally arrive at the derivation of exact Euclidean coordinates for both rows and columns of the contingency table in common space called *dual space* in Chap. 5.

The key points will be fully explained and illustrated using a number of simple numerical examples. It is hoped that the readers will see the core of the procedure and agree that there is nothing fundamentally wrong with the proposed procedure. In spite of the enormous popularity of the current methods of joint graphical display (symmetric scaling and non-symmetric scaling), we will show that they are far from being perfect.

The second challenge is what we can do to promote data analysis after the successful derivation of exact Euclidean coordinates for both row and column variables of the contingency table. The fact remains that we are not typically capable of graphing data structure of more than three dimensions. We need either to develop a practical and dynamic graphical method for multidimensional space or else some useful alternatives to graphical display.

In concluding this chapter, therefore, the readers should know the following points about joint graphical display:

- The currently popular graphical method of symmetric scaling (French plot) offers an approximation to the correct graph, hence it is not perfect as many readers may believe.
- Symmetric scaling is a practical compromise as we will see later in Chap. 5. In this regard, we should salute the French wisdom for the brilliant compromise to swim away from the practical difficulty.
- The currently popular graphical method of non-symmetric scaling is logically incorrect.
- The proposed method for graphing in doubled multidimensional space provides correct Euclidean coordinates for both rows and columns in common space, hence a solution to our perennial problem of joint graphical display.
- In this derivation, we will introduce a new framework for multidimensional space by defining *contingency space, dual space, residual space* and *total space*. This will forge a new framework for doubled multidimensional space and quantification theory.

- From the practical point of view, we will be left with the question of how to deal with graphs of more than three dimensions.
- A drastic departure from joint graphical display has already been initiated by applications of cluster analysis, as we will see in Chap. 6. We will discuss the question if this alternative is a good compromise for exact multidimensional joint graphs.
- Thus, solving the perennial problem of joint graphical display is far from concluding the quest for how to represent multidimensional relations in categorical data.

To end this reminiscence, the following points are noted:

1. R. Darrell Bock, known for his optimal scaling (Bock 1960), was the supervisor of Nishisato's PhD thesis, entitled "Minimum entropy clustering of test items" (University of North Carolina at Chapel Hill 1966). So, after many years of work on optimal scaling and dual scaling, he is now returning to the starting point of his career, that is, cluster analysis.
2. Nishisato's lifework over 60 years was dominated by "transformations" of data so as to accommodate different purposes of analysis. The work started with changing the traditional unidimensional analysis (e.g., Taylor's manifest anxiety scale 1953) to multidimensional analysis (Nishisato 1960), factor analysis of time series data (Nishisato 1965, 1970a), logistic regression of binary data for clustering (Nishisato 1966, 1970b), transform factor analysis (Nishisato 1971a), analysis of variance of categorical data (Nishisato 1971b), standardization effects (Nishisato and Yamauchi 1974), data formats versus product-moment, polychoric and canonical correlation (Nishisato and Torii 1971; Nishisato and Hemsworth 2002), nonlinear programming for scaling (Nishisato and Arri 1975), scaling of different types of categorical data (Nishisato 1978, 1980, 1993, 1996; Nishisato and Nishisato 1984), piecewise method of reciprocal averages, *an extension of its original analysis of contingency tables to multiple-choice data* (Nishisato and Sheu 1980), method of reciprocal medians for robust analysis (Nishisato 1984b, 1987), projections of data onto target subspaces (Nishisato 1984a, 1986a, 1988a, b; Nishisato and Gaul 1990; Nishisato and Baba 1999), decision rules for when not to analyze data (Nishisato and Ahn 1995), scaling of multi-way data (Nishisato and Lawrence 1989), standardizing multidimensional space (Nishisato 1991), cluster analysis through filters (Nishisato 2014), and bridging between quantification and clustering (Nishisato 2020a, b; Nishisato and Clavel 2003, 2010; Clavel and Nishisato 2012, 2020). His lifework thus expanded applications of data analysis to a wider variety of problems through "focusing, projection and transformation of data" as the central role.

In Chap. 6, Clavel and Nishisato will investigate applications of cluster analysis as an alternative to joint graphical display, through many numerical examples. The two authors will then finish Part 1 with some further thoughts on graphical and clustering approaches to quantification theory. Clavel has played an indispensable role as Nishisato's collaborator for over the past twenty years by carrying out most numerical analyses for him. Clavel is also instrumental to the idea of "clustering of modified data matrices (CMDM)" as presented in Chap. 6.

Before moving to the core of the problem in Chaps. 3–5, let us first indulge in some preliminary mathematical background for graphical display in Chap. 2.

References

Adachi, K. (2006). *Tahenryo Data Kaiseki Ho (Multivariate Data Analysis)*. Tokyo: Nakanishiya Publisher.

Adachi, K., & Murakami, T. (2011). *Hikeiryou Tahenryou Kaisekihou: Shseibun Bunseki kara Tajyuu Taiou Bunseki e (Nonmetric Multivariate Analysis: From Principal Component Analysis to Multiple Correspondence Analysis)*. Tokyo: Asakura Shoten.

Akiyama, S. (1993). *Suryouka no Graphics: Taido no Tahenryoukaiseki (Graphics for Quantification: Multivariate Analysis of Attitudes)*. Tokyo: Asakura Shoten.

Beh, E. J. (2004). *A Bibliography of the Theory and Applications of Correspondence Analysis*.

Beh, E. J., & Lombardo, R. (2014). *Correspondence Analysis: Theory, Practice and New Strategies*. United Kingdom: Wiley.

Beltrami, E. (1873). Sulle funzioni bilineari (On the bilinear functions). In G. Battagline & E. Fergola (Eds.), *Giornale di Mathematiche, 11* (pp. 98–106).

Benzécri, J. P. (1992). *Correspondence Analysis Handbook*. New York: Marcel Dekker.

Benzécri, J. P., et al. (1973). *L'analyse des données: II. L'analyse des correspondances*. Paris: Dunod.

Birks, H. J. B., Peglar, S. M., & Austin, H. A. (1996). An annotated bibliography of canonical correspondence analysis and related constrained ordination methods. *Abstracta Botanica, 20,* 17–36.

Blasius, J., & Greenacre, M. J. (1998). *Visualization of Categorical Data*. London: Academic Press.

Bock, R. D. (1956). The selection of judges for preference testing. *Psychometrika, 21,* 349–366.

Bock, R. D. (1960). Methods and applications of optimal scaling. *The University of North Carolina Psychometric Laboratory Research Memorandum*, No: 25.

Bouroche, J. M. (1977). *Analyse des Données en Marketing*. Paris: Masson.

Bouroche, J. M., & Saporta, G. (1980). *L'Analyse des Données*. Paris: Presses Univessitaires de France.

Caillez, F., & Pagés, J. P. (1976). *Introduction a L'Analyse des Données*. Paris: SMASH.

Carroll, J. D., Green, P. E., & Schaffer, C. M. (1986). Interpoint distance comparisons in correspondence analysis. *Journal of Marketing Research, 23,* 271–280.

Carroll, J. D., Green, P. E., & Schaffer, C. M. (1987). Comparing interpoint distances in correspondence analysis: A clarification. *Journal of Marketing Research, 24,* 445–450.

Carroll, J. D., Green, P. E., & Schaffer, C. M. (1989). Reply to Greenacre's commentary on the Carroll-Green-Schaffer scaling of two-way correspondence analysis solutions. *Journal of Marketing Research, 26,* 366–368.

Cibois, P. (1983). *L'Analyse Factorielle*. Paris: Presses Universitaires de France.

Clausen, S. E. (1998). *Applied Correspondence Analysis: An Introduction*. Sage Publications.

Clavel, J. G., & Nishisato, S. (2012). Reduced versus complete space configurations in total information analysis. In W. Gaul, A. Geyer-Schultz, Schmidt-Tiéme, & J. Kunze (Eds.), *Challenges at the Interface of Data Analysis Computer-Science and Optimization* (pp. 91–99). Springer.

Clavel, J. .G., & Nishisato, S., et al. (2020). From joint graphical display to bi-modal clustering: [2] Dual space versus total space. In T. Imaizumi (Ed.), *Advanced Research in Classification and Data Science* (pp. 131–143). Singapore: Springer.

de Lagarde, J. (1983). *Initiation á l'Analyse des Données*. Paris: Dunod.

de Leeuw, J. (1984). *Canonical Analysis of Categorical Data*. Leiden University: DSWO Press.

Eckart, C., & Young, G. (1936). The approximation of one matrix by another of lower rank. *Psychometrika, 1,* 211–218.
Edgerton, H. A., & Kolbe, L. E. (1936). The method of minimum variation for the composite criteria. *Psychometrika, 1,* 183–187.
Escofier, B., & Pagés, J. (1988). *Analyses Factorielles Simples et Multiples*. Paris: Dunod.
Fisher, R. A. (1940). The precision of discriminant functions. *Annals of Eugenics, 10,* 422–429.
Fisher, R. A. (1948). *Statistical methods for research workers*. London: Oliver and Boyd.
Gabriel, K. R. (1971). The biplot graphical display of matrices with applications to principal component analysis. *Biometrics, 58,* 453–467.
Gauch, H. G. (1982). *Multivariate Analysis in Community Ecology*. Cambridge: Cambridge University Press.
Gifi, A. (1990). *Nonlinear Multivariate Analysis*. New York: Wiley.
Gower, J. C., & Hand, D. J. (1996). *Biplots*. London: Chapman & Hall.
Greenacre, M. J. (1984). *Theory and Applications of Correspondence Analysis*. London: Academic Press.
Greenacre, M. J. (1989). The Carroll-Green-Schaffer scaling in correspondence analysis: A theoretical and empirical appraisal. *Journal of Marketing Research, 26,* 358–365.
Greenacre, M. J. (1993). *Correspondence Analysis in Practice*. London: Academic Press.
Greenacre, M. J., & Blasius, J. (Eds.). (1994). *Correspondence Analysis in the Social Sciences*. London: Academic Press.
Greenacre, M. J., & Blasius, J. (Eds.). (2006). *Multiple Correspondence Analysis and Related Methods*. Boca Raton: Chapman and Hall/CRC.
Guttman, L. (1941). The quantification of a class of attributes: A theory and method of scale construction. In The Committee on Social Adjustment (Ed.), *The Prediction of Personal; Adjustment* (pp. 319–348). New York: Social Science Research Council.
Guttman, L. (1946). An approach for quantifying paired comparisons and rank order. *Annals of Mathematical Statistics, 17,* 144–163.
Hayashi, C. (1950). On the quantification of qualitative data from the mathematico-statistical point of view. *Annals of the Institute of Statistical Mathematics, 2,* 35–47.
Hayashi, C. (1952). On the prediction of phenomena from qualitative data and the quantification of qualitative data from the mathematico-statistical point of view. *Annals of the Institute of Statistical Mathematics, 3,* 69–98.
Hayashi, C. (1974). *Suryouka no Houhou (Methods of Quantification)*. Tokyo: Tokyo Keizai Sha.
Hayashi, C. (1993). *Suryouka: Riron to Hoho (Quantification: Theory and Methods)*. Tokyo: Asakura Shoten.
Hayashi, C. (2001). *Data no Kagaku (Data Science)*. Tokyo: Asakura Shoten.
Hayashi, C., Higuchi, I., & Komazawa, T. (1970). *Johoshori to Tokeisuiri (Information Processing and Statistical Mathematics)*. Tokyo: Sangyou Tosho.
Hayashi, C., & Suzuki, T. (1986). *Shakai Chosa to Suryouka (Social Surveys and Quantification)*. Tokyo: Iwanami Shoten.
Hill, M. O. (1973). Reciprocal averaging: An eigenvector method of ordination. *Journal of Ecology, 61,* 237–249.
Hill, M. O. (1974). Correspondence analysis: a neglected multivariate method. *Journal of the Royal Statistical Society C (Applied Statistics), 23,* 340–354.
Hirschfeld, H. O. (1935). A connection between correlation and contingency. *Cambridge Philosophical Society Proceedings, 31,* 520–524.
Horst, P. (1935). Measuring complex attitudes. *Journal of Social Psychology, 6,* 369–374.
Horst, P. (1936). Obtaining a composite measure from a number of different measures of the same attribute. *Psychometrika, 1,* 53–60.
Hotelling, H. (1933). Analysis of complex of statistical variables into principal components. *Journal of Educational Psychology*, *24*, 417–441 and 498–520.
Iwatsubo, S. (1987). *Suryuka no Kiso (Foundations of Quantification)*. Tokyo: Asakura Shoten.
Jambu, M. (1989). *Exploration Informatique et Statistique de Données*. Paris: Dunod.

Jambu, M., & Lebeaux, M. O. (1978). *Classification Automatique pour l'Analyse des Données. Métodes et Algorithmes*. Paris: Dunod.

Johnson, P. O. (1950). The quantification of qualitative data in discriminant analysis. *Journal of the American Statistical Association, 45,* 65–76.

Johnson, R. M. (1963). On a theorem stated by Eckart and Young. *Psychometrika, 28,* 259–263.

Jordan, C. (1874). Mémoire sur les formes bilinieres (Note on bilinear forms). *Journal de Mathématiques Pures et Appliquées, deuxiéme Série, 19,* 35–54.

Kiers, H. (1989). *Three-Way Methods for the Analysis of Qualitative and Quantitative Two-Way Data*. Leiden University: DSWO Press.

Komazawa, T. (1978). *Tagenteki Data Bunseki no Kiso (Foundations of Multidimensional Data Analysis)*. Tokyo: Asakura Shoten (in Japanese).

Komazawa, T. (1982). *Suryoka Riron to Data Shori (Quantification Theory and Data Analysis)*. Tokyo: Asakura Shoten (in Japanese).

Komazawa, T., Higuchi, I., & Ishizaki, R. (1998). *Pasokon Suryoukabunseki (Quantification Anlysis with Personal Computer)s*. Tokyo: Asakura Shoten.

Koster, J. T. A. (1989). *Mathematical Aspects of Multiple Correspondence Analysis for Ordinal Variables*. Leiden University: DSWO Press.

Lebart, L., Morineau, A., & Fénelon, J. P. (1979). *Traitement des Données Statistiques*. Paris: Dunod.

Lebart, L., Morineau, A., & Tabard, N. (1977). *Techniques de la Description Statistique: Méthodes et Logiciels pour l'Analyse des Grands Tableaux*. Paris: Dunod.

Lebart, L., Morineau, A., & Warwick, K. M. (1984). *Multivariate Descriptive Statistical Analysis*. New York: Wiley.

Lefevre, J. (1976). *Introduction aux Analyses Statistiques Multidimensionnelles*. Paris: Masson.

Le Roux, B., & Rouanet, H. (2004). *Geometric Data Analysis: From Correspondence Analysis to Structured Data*. Dordrecht: Kluwer.

Lingoes, J. C. (1978). *Geometric Representation of Relational Data*. Ann Arbor: Mathesis Press.

Lingoes, J. C., Roskam, E. E., & Borg, I. (1978). *Geometric Representation of Relational Data*. Ann Arbor: Mathesis Press.

Lord, F. M. (1958). Some relations between Guttman's principal components of scale analysis and other psychometric theory. *Psychometrika, 23,* 291–296.

Maung, K. (1941). Measurement of association in contingency tables with special reference to the pigmentation of hair and eye colours of Scottish children. *Annals of Eugenics, 11,* 189–223.

McDonald, R. P. (1968). A unified treatment of the weighting problem. *Psychometrika, 33,* 351–381.

McKeon, J. J. (1966). Canonical analysis: some relations between canonical correlation, factor analysis, discriminant function analysis and scaling theory. *Psychometric Monograph No. 13*.

Meulman, J. (1982). *Homogeniety Analysis of Incomplete Data*. Leiden University: DSWO Press.

Michailidis, G., & de Leeuw, J. (1998). The Gifi system of descriptive multivariate analysis. *Statistical Science, 13,* 307–336.

Moreau, J., Doudin, P. A., & Cazes, P. (1999). *L'Analyse des correspondances et les Techniques Connexes: Approches Nouvelles pour l'Analyse Statistique des Données*. Berlin: Springer.

Mosier, C. .I. (1946). Machine methods in scaling by reciprocal averages. *Proceedings, Research Forum* (pp. 35–39). Endicath, N.Y.: International Business Corporation.

Murtagh, F. (2005). *Correspondence Analysis and Data Coding with R and Java*. Boca Raton: Chapman and Hall.

Nakache, J. P. (1982). *Exercices Comment'es de Mathématiques pour l'Analyse des Données*. Paris: Dunod.

Nishisato, S. (1960). Factor analytic study of an anxiety scale. *Japanese Journal of Psychology, 35,* 228–236 (in Japanese).

Nishisato, S. (1965). A simple method of time series analysis. Memorial Volume of Professor Kin-ichi Yuki,103-111. Sapporo: Yamafuji Press] (in Japanese).

Nishisato, S. (1966). *Minimum entropy clustering of test items*. Ph.D. Thesis, the University of North Carolina at Chapel Hill, N.C., USA.

Nishisato, S. (1970a). Factor analysis of time series – a single stationary case. In *Proceedings of the 34th Annual Conference of the Japanese Psychologiccal Association*, Sendai, 97. (in Japanese)
Nishisato, S. (1970b). Probability estimation of dichotomous response patterns by logistic fractional-factorial representation. *Japanese Psychological Research, 12,* 87–95.
Nishisato, S. (1971a). Transform factor analysis: A sketchy presentation of a general approach. *Japanese Psychological Research, 13,* 155–166.
Nishisato, S. (1971b). Analysis of variance through optimal scaling. In *Proceedings of the First Canadian Conference in Applied Statistics* (pp. 306–316). Montreal: Sir George Williams University Press.
Nishisato, S. (1972). Optimal Scaling and Its Generalizations. I. Methods. Department of Measurement and Evaluation, OISE: Toronto.
Nishisato, S. (1973). Optimal Scaling and Its Generalizations. II. Applications. Department of Measurement and Evaluation, OISE: Toronto.
Nishisato, S. (1975). *Oyo Shinri Shakudoho: Shitsuteki Data no Bunseki to Kaishaku (Applied Psychological Scaling: Analysis and Interpretation of Qualitative Data)*. Tokyo: Seishin Shobo (in Japanese).
Nishisato, S. (1976). *Optimal Scaling as Applied to Different Forms of Categorical Data*. Department of Measurement and Evaluation, OISE: Toronto.
Nishisato, S. (1978). Optimal scaling of paired comparison and rank order data: An alternative to Guttman's formulation. *Psychometrika, 43,* 267–271.
Nishisato, S. (1979). *An Introduction to Dual Scaling*. Department of Measurement and Evaluation, OISE: Toronto.
Nishisato, S. (1980). *Analysis of Categorical Data: Dual Scaling and Its Applications*. Toronto: The University of Toronto Press.
Nishisato, S. (1982). *Shitsuteki Data no Suryouka: Soutsuishakudo-ho to Sono Ohyou (Quantification of Qaulittative Data: Dual Scaling and its Applications)*. Tokyo: Asakura Shoten.
Nishisato, S. (1984a). Forced classification: A simple application of a quantification technique. *Psychometrika, 49,* 25–36.
Nishisato, S. (1984b). Dual scaling by reciprocal medians. In *Proceedings of the 32nd Scientific Conference of the Italian Statistical Society*. (pp. 141–147) Italy: Sorrento.
Nishisato, S. (1986a). Generalized forced classification for quantifying categorical data. In E. Diday (Ed.), *Data Analysis and Informatics* (pp. 351–362). Amsterdam: North-Holland.
Nishisato, S. (1986b). *Quantification of Categorical Data: A Bibliography 1975–1986*. Toronto: MicroStats.
Nishisato, S. (1987). Robust techniques for quantifying categorical data. In I. B. MacNeil & G. J. Umphrey (Eds.), *Foundations of Statistical Inference, 209–217*. Dordrecht, The Netherlands: D. Reidel Publishing Company.
Nishisato, S. (1988a). Forced classification procedure of dual scaling: its mathematical properties. In H. H. Bock (Ed.), *Classificaiton and Related Methods* (pp. 523–532). Amsterdam: North-Holland.
Nishisato, S. (1988b). Market segmentation by dual scaling through generalized forced classification. In W. Gaul & M. Schader (Eds.), *Data, Expert Knowledge and Decisions* (pp. 268–278). Berlin: Springer.
Nishisato, S. (1991). Standardizing multidimensional space for dual scaling. In *Proceedings of the 20th Annual Meeting of the German Operations Research Society* (pp. 584–591). Hohenheim University.
Nishisato, S. (1993). On quantifying different types of categorical data. *Psychometrika, 58,* 617–629.
Nishisato, S. (1994). *Elements of Dual Scaling: An Introduction to Practical Data Analysis*. Hilsdale, N.J.: Lawrence Erlbaum Associates.
Nishisato, S. (1996). Gleaning in the field of dual scaling. *Psychometrika, 61,* 559–599.
Nishisato, S. (2007a). *Multidimensional Nonlinear Descriptive Analysis*. London: Chapman-Hall/CRC.

Nishisato, S. (2007b). *Insight into Data Analysis: Fundamentals of Quantification*. Nishinomiya: Kwansei Gakuin University Press (in Japanese).
Nishisato, S. (2010). *Data Analysis for Behavioral Sciences: Use of Methods Appropriate for Information Retrieval*. Tokyo: Baifukan (in Japanese).
Nishisato S. (2012). Quantification theor: Reminiscence and a step forward. In W. Gaul, A. Geyer-Schultz, Schmidt-Tiéme, & J. Kunze (Eds.), *Challenges at the Interface of Data Analysis, Computer-Science and Optimization* (pp. 109–119). Springer.
Nishisato, S. (2014). Structural representation of categorical data and cluster analysis through filters. In W. Gaul, A. Geyer-Schultz, Y. Baba, & A. Okada (Eds.), *German-Japanese Interchange of Data Analysis Results* (pp. 81–90). Springer.
Nishisato, S. (2016). Quantification theory: Dual space and total space. Paper presented at the annual meeting of the Behaviormetric Society, Sapporo, Japan, p. 27 (In Japanese)
Nishisato, S. (2019a). Reminiscence: Quantification theory and graphs. *Theory and Applications of Data Analysis, 8,* 47–57 (in Japanese).
Nishisato, S. (2019b). Expansion of contingency space: Theory of doubled multidimensional space and graphs. An invited talk at the Annual Meeting of the Japanese Classification Society, Tokyo (in Japanese).
Nishisato, S. (2020a). From joint graphical display to bi-modal clustering: [1] A giant leap in quantification theory. In T. Imaizumi (Ed.), *Advanced Research in Classification and Data Science* (pp. 157–168). Singapore: Springer.
Nishisato, S. (2020b). Quantification theory: Categories, variables and mode of analysis. In T. Imaizumi (Ed.), *Advanced Studies in Behaviormetrics and Data Science* (pp. 253–264). Singapore: Springer.
Nishisato, S., & Ahn, H. (1995). When not to analyze data: Decision making on missing responses in dual scaling. *Annals of Operations Research, 55,* 361–378.
Nishisato, S., & Arri, P. S. (1975). Nonlinear programming approach of optimal scaling to partially ordered categories. *Psychometrika, 40,* 525–548.
Nishisato, S., & Baba, Y. (1999). On contingency, projection and forced classification of dual scaling. *Behaviormetrika, 26,* 207–219.
Nishisato, S., & Clavel, J. G. (2003). A note on between-set distances in dual scaling and correspondence analysis. *Behaviormetrika, 30,* 87–98.
Nishisato, S., & Clavel, J. G. (2010). Total information analysis: Comprehensive dual scaling. *Behaviormetrika, 37,* 15–32.
Nishisato, S., & Gaul, W. (1990). An approach to marketing data analysis: The forced classificaion procedure of dual scaling. *Journal of Marketing Resarch, 27,* 354–360.
Nishisato, S., & Hemsworth, D. (2002). Quantification of ordinal variables: A critical inquiry into polychoric and canonical correlation. In Y. Baba, A. J. Hayter, K. Kanefuji, & S. Kuriki (Eds.), *Recent Advances in Statistical Research and Data Analysis* (pp. 49–84). Tokyo: Springer.
Nishisato, S., & Lawrence, D. R. (1989). Dual scaling of multiway data matrices: several variants. In R. Coppi & S. Bolasco (Eds.), *Multiway Data Analysis* (pp. 317–326). Amsterdam: Elsevier Science Publishers.
Nishisato, S., & Leong, K. S. (1975). *OPSCAL: A Fortran IV Program for Analysis of Qualitative Data by Optimal Scaling*. Department of Measurement and Evaluation, OISE: Toronto.
Nishisato, S., & Nishisato, I. (1986). *Dual3 Users' Guide*. Toronto: MicroStats.
Nishisato, S., & Nishisato, I. (1984). *An Introduction to Dual Scaling*. Toronto: MicroStats.
Nishisato, S., & Nishisato, I. (1994). *Dual Scaling in a Nutshell*. Toronto: MicroStats.
Nishisato, S., & Sheu, W. (1980). Piecewise method of reciprocal averages for multiple-choice data of multiple-choice data. *Psyhometrika, 45,* 467–478.
Nishisato, S., & Torii, Y. (1971). Assessment of information loss in scoring monotone items. *Multivariate Behavioral Research, 6,* 91–103.
Nishisato, S., & Yamauchi, H. (1974). Principal components of deviation scores and standard scores. *Japanese Psychological Research, 16,* 162–170.

Ohsumi, N., Lebart, L., Morineau, A., Warwick, K. M., & Baba, Y. (1994). *Kijyutsuteki Tahenryou Kaiseki (Descriptive Multivariate Analysis)*. Tokyo: Nikkagiren (in Japanese).
Pearson, K. (1901). On lines and planes of closest fit to systems of points in space. *Philosophical Magazines and Journal of Science, Series, 6*(2), 559–572.
Richardson, M., & Kuder, G. F. (1933). Making a rating scale that measures. *Personnel Journal, 12,* 36–40.
Saito, T. (1980). *Tajigen Shakudo Kouseihou (Multidimensional Scale Construction)*. Tokyo: Asakura Shoten.
Saporta, G. (1979). *Theories et Méthodes de la Statistique*. Paris: Technip.
Schmidt, E. (1907). Zür Theorie der linearen und nichtlinearen Integralgleichungen. Esster Teil. Entwickelung willkürlicher Functionen nach Systemaen vorgeschriebener (On theory of linear and nonlinear integral equations. Part one. Development of arbitrary functions according to prescribed systems). *Mathematische Annalen, 63,* 433–476.
Schönemann, Bock, R. D. & Tucker, L. R. (1965). Some notes on a theorem by Eckart and Young. *Research Memorandum, No.25*. The Psychometric Laboratory, University of North Carolina.
Slater, P. (1960). Analysis of personal preferences. *British Journal of Statistical Psychology, 3,* 119–135.
Takane, Y. (1980). *Tajigen Shakudo Ho (Multidimensional Scaling)*. Tokyo: University of Tokyo Press (in Japanese).
Takane, Y. (1995). *Seiyakutsuki Shuseibun Bunsekiho: Atarashii Tahenryou Kaisekiho (Constrained Principal Component Analysis: A New Approach to Multivariate Data Analysis)*. Tokyo: Asakura Shoten.
Takane, Y., Yanai, H., & Mayekawa, S. (1991). Relationships among several methods of linearly constrained correspondence analysis. *Psychometrika, 56,* 667–684.
Takeuchi, K., & Yanai, H. (1972). *Tahenryou Kaiseki no Kiso (Foundations of Multivariate Analysis)*. Tokyo: Toyo Keizai-sha (in Japanese).
Taylor, J. (1953). A personality scale of manifest anxiety. *The Journal of Abnormal and Social Psychology, 48,* 285–290.
Teil, H. (1975). Correspondence factor analysis: An outline of its method. *Mathematical Geology, 7,* 3–12.
Tenenhaus, M. (1982). *Multiple correspondence analysis and duality schema: a synthesis of different approaches* (pp. 289–302). XL: Metron.
Tenenhaus, M. (1994). *Méthodes Statistiques en Gestion*. Paris: Dunod.
Tenenhaus, M., & Young, F. W. (1985). An analysis and synthesis of multiple correspondence analysis, optimal scaling, dual scaling, homogeneity analysis and other methods for quantifying categorical data. *Psychometrika, 50,* 91–119.
van Buuren, S. (1990). *Optimal Scaling of Time Series*. Leiden University: DSWO Press.
van de Geer, J. P. (1993). *Multivariate Analysis of Categorical Data: Applications*. Newbury Park: Sage.
van der Burg, E. (1988). *Nonlinear Canonical Correlation and Some Related Techniques*. Leiden University: DSWO Press.
van der Heijden, P. G. M. (1987). *Correspondence Analysis of Longitudinal Data*. Leiden University: DSWO Press.
van Rijckevorsel, J. (1987). *The Applications of Fuzzy Coding and Horseshoes in Multiple Correspondence Analysis*. Leiden University: DSWO Press.
Verboon, P. (1994). *A Robust Approach to Nonlinear Multivariate Analysis*. Leiden University: DSWO Press.
Weller, J. C., & Romney, A. K. (1990). *Metric Scaling: Correspondence Analysis*. Newbury: Sage Publications.
Whittaker, R. H. (1978). *Ordination of Plant Communities*. The Hague: Junk.
Wilks, S. S. (1938). Weighting system for linear function of correlated variables when there is no dependent variable. *Psychometrika, 3,* 23–40.

Williams, E. J. (1952). Use of scores for the analysis of association in contingency tables. *Biometrika, 39,* 274–289.

Yanai, H., & Takane, Y. (1977). *Tahenryou Kaiseki Ho (Multivariate Analysis).* Tokyo: Asakura Shoten.

Yanai, H., & Takagi, H. (1986). *Teneryoukaiseki Handbook (Handbook of Multivariate Analysis).* Tokyo: Gendai Sugaku Sha.

Young, F. W., de Leeuw, J., & Takane, Y. (1976). Regression with qualitative and quantitative variables: an alternating least squares method with optimal scaling features. *Psychometrika, 41,* 505–529.

Young, F. W., Takane, Y., & de Leeuw, J. (1978). The principal components of mixed measurement level multivariate data: an alternating least squares method with optimal scaling features. *Psychometrika, 43,* 279–281.

Chapter 2
Mathematical Preliminaries

This combination is logically confusing, but it is the usual way that our students are taught. Should they not be taught that the correlation would influence the shape of the graph and the orientation of the axes? Can we assume that the students know how the correlation must be incorporated into the graph or how to draw a graph of correlated test scores in the orthogonal coordinate system?

Redundant as it may be, let us discuss how to draw a graph, using orthogonal axes, namely one horizontal and the other vertical axes, to represent students' scores on two correlated tests (mathematics score, English score). Here, one orthogonal axis is no longer the axis for the mathematics test and the other orthogonal axis is no longer the axis for the English test. This tiny step is a giant leap from the old-fashioned illogical way to the sound mathematical way.

To express students' scores on the two tests, say (x_{1i}, x_{2i}) for student i, it is much more expeditious to start with orthogonal axes than with oblique axes, where "obliqueness" is introduced by the correlation between the two tests: if the correlation is 1, we need a single axis; when the correlation is 0, we need two axes with the angle of 90 degrees. Then, our question is how to introduce an orthogonal coordinate system for paired scores of students with the correlation somewhere between 0 and 1. This is an important step towards the correct geometric graph.

To achieve the above task, we follow essentially the same procedure as used in the principal component analysis, PCA (Pearson 1901; Hotelling 1933). This introduction will help us later since the topic of this book, quantification theory (QT), is called *principal component analysis (PCA) of categorical data* (Torgerson 1958). Thus, this discussion will be useful as a preview of quantification theory.

S. Nishisato et al., *Modern Quantification Theory*,
Behaviormetrics: Quantitative Approaches to Human Behavior 8,
https://doi.org/10.1007/978-981-16-2470-4_2

2.1 Graphs with Orthogonal Coordinates

Let us consider two sets of achievement scores, one for mathematics and the other for English, and suppose that we obtained those scores from N students. Let us start with the traditional way of graphing the data with the horizontal axis for the mathematics test and the vertical axis for the English test.

This familiar graph typically may show us such tendency that the higher the scores on X_1, the higher the scores on X_2. Such a graph indicates that the two tests are positively correlated. To be more precise, we can calculate the correlation by the formula

$$r_{12} = \sqrt{\frac{\sum_{k=1}^{N}(X_{1k} - m_1)(X_{2k} - m_2)}{Ns_1 s_2}},$$

where m_1 and m_2 are means of the Mathematics test and the English test, respectively, and s_1 and s_2 are standard deviations of the two tests.

Since the two tests are typically correlated, it is logical to assume that we cannot use orthogonal coordinates to represent the two tests. We will see shortly that if the two tests are correlated by r_{12}, the geometric representation of the two tests can be represented by two axes with the angle equal to $cos^{-1} r_{12}$. The question is how to introduce such axes in representing the two tests.

2.1.1 *Linear Combination of Variables*

To incorporate this knowledge to arrive at a mathematically correct graph for two correlated tests, we will consider a linear combination of the two variables, defined by

$$Y_i = \beta_1 X_{1i} + \beta_2 X_{2i}$$

for Subject i, under the condition that

$$\beta_1{}^2 + \beta_2{}^2 = 1.$$

In this expression, β_1 and β_2 are called weights for the mathematics test and the English test, respectively, in our example. This general form of the linear combination has an important aspect that we should know:

- The condition that $\beta_1{}^2 + \beta_2{}^2 = 1$ assures us that the combined score, called a linear combination of the two variables or composite score, is defined as the projection of two sets of scores onto the composite score axis Y.

How this composite axis is decided depends on the purpose of the analysis as we will shortly discuss it. The underlying rationale for starting with the linear combination

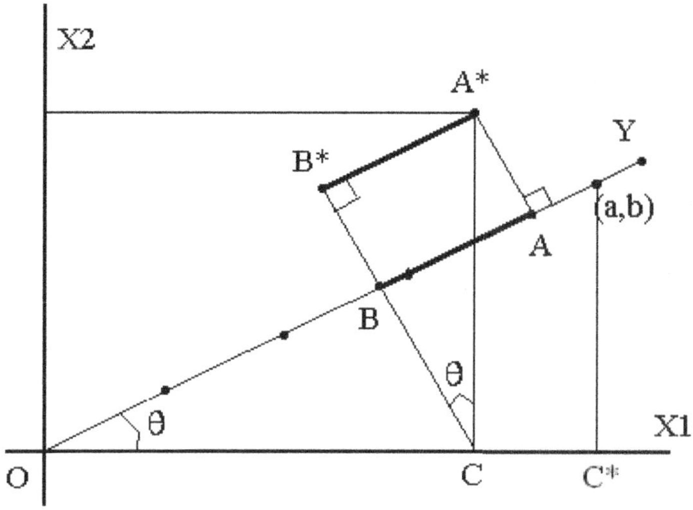

Fig. 2.1 Calculation of projection of a point onto a chosen axis

to arrive at the first composite axis is to introduce the notion of *projection* and then orthogonal axes.

Once we have the first axis used for the projection of data on it, then we can introduce another axis for the linear combination of the two variables in such a way that the second axis is orthogonal to the first axis. The simplest procedure to achieve this task is that once we obtain the first linear combination of variables, we calculate the residual scores (i.e., the original scores minus the first composite scores), and then consider other linear combinations of the residual scores, which will yield the second composite scores. The projection of the residual scores on another axis assures that it is orthogonal to the first composite scores. We can continue the same process to extract more orthogonal components until no more residual scores can be found.

This procedure can easily be extended to a linear combination of more than two variables, and repeat essentially the same procedure to arrive at a set of scores projected onto orthogonal axes for many variables. This is a basic procedure to generate an orthogonal coordinate system for many correlated variables in the Euclidean space.

Since the condition that the sum of squares of weights be 1 is a basis for our discussion, let us illustrate this point using a small graphical scheme (Fig. 2.1).

Consider two orthogonal axes (X_1, X_2) and introduce an arbitrary axis Y, which goes through the origin O and the point (a, b). Let us calculate the projection of an arbitrary point A^* with the coordinate (x_1, x_2), that is, $A^* : (x_1, x_2)$, onto axis Y, and indicate the projected point by A. Then

$$\overline{OA} = Y^* = \overline{OB} + \overline{BA} = \overline{OB} + \overline{B^*A^*} = x_1 \cos\theta + x_2 \sin\theta = \beta_1 x_1 + \beta_2 x_2.$$

Note that

$$\cos\theta = \frac{a}{\sqrt{a^2+b^2}} = \beta_1, \qquad \sin\theta = \frac{b}{\sqrt{a^2+b^2}} = \beta_2,$$

hence

$$\beta_1^2 + \beta_2^2 = 1.$$

In other words, the sum of squares of the weights (β_1 and β_2) is 1. This condition is known to make certain that the composite score and the *projection* of the original scores onto the chosen composite axis Y have the same unit.

If we have n variables (e.g., tests), one can consider a composite score given by

$$Y = \beta_1 x_1 + \beta_2 x_2 + \beta_3 x_3 + \cdots + \beta_n x_n = \sum\nolimits_{i=1}^{n} \beta_i x_i,$$

where

$$\sum\nolimits_{i=1}^{n} \beta_i^2 = 1.$$

2.1.2 Principal Axes

Now that we understand how to project a score onto an arbitrary axis, let us ask the following question: "What is the most meaningful way to choose Y?" This is important because this question of "what is" determines the nature of the composite scores.

The *most popular choice* is that the composite score, which is unidimensional (i.e., one can express it on a single axis), should be as representative as possible of the sets of scores (i.e., data), scattered in two-dimensional space (Note: in the general case, this will be multidimensional space). This popular choice can be attained by choosing the weights in such a way that *the composite score has the maximal variance*. In other words, such a composite axis must be that all the data points are as close as possible to the composite axis. When this is done, the resultant axis Y is called the *principal axis*. Thus, the well-known principal component analysis (PCA) (see Pearson 1901; Hotelling 1933) can be rephrased as follows:

- Consider a linear combination of n variables in such a way that the variance of the composite scores has the maximal variance, subject to the condition that the sum of the squared weights is equal to 1.

This normalizing condition means mathematically that we are projecting a set of variables onto the chosen axis of the same unit. When we obtain such an axis that maximizes the variance of the linear combination, the axis is called the *principal axis*. The resultant composite scores are called the first *principal component scores*.

In formulating PCA, we typically differentiate the variance-covariance matrix of the composite scores with respect to the weights, under the normalizing condition, set the derivatives equal to zero, and solve it for the weights. This process leads to the eigenequation, and we know how to solve it. For the detailed derivation of the eigenequation, please refer to any of the standard books on multivariate analysis (e.g., Rao 1952; Mardia et al. 1979; Takeuchi et al. 1982).

Once the first principal component scores are obtained, we calculate the residual scores (i.e., the observed scores minus principal component scores for each variable). We then consider a linear combination of the residual scores that maximize the variance to arrive at the second principal component. By repeating the same process, we finally reach the stage where all the variance of the n variables is exhaustively accounted for and we have a set of (orthogonal) principal coordinates, together with the eigenvalues (the variance of principal components).

2.2 Correlation and Orthogonal Axes

Let us now consider only two variables, which we indicate as $X_{1i}, X_{2i}, i = 1, 2, \ldots, N$ (e.g., two sets of scores from N subjects). Let us indicate the two means of the test by m_j, $j = 1, 2$. Then, the variance of set j, ${s_j}^2$, $j = 1, 2$, and covariance s_{12} can be written as

$$ {s_j}^2 = \frac{\sum_{i=1}^{N}(X_{ij} - m_j)^2}{N-1}, \qquad s_{12} = \frac{\sum_{i=1}^{N}(X_{i1} - m_1)(X_{i2} - m_2)}{N-1}. $$

Then, the product-moment correlation between the two tests, r_{12}, is given by

$$ r_{12} = \frac{s_{12}}{s_1 s_2}. $$

In linear algebra, it is well known that once we express two sets of scores as two vectors, say $\mathbf{x_1}$, $\mathbf{x_2}$, the product-moment correlation can be expressed as the cosine of the angle between the two vectors (e.g., see Takeuchi et al. (1982) for its derivation and proof) that

$$ r_{12} = cos\theta_{12}, $$

where θ_{12} is the angle between the two vectors $\mathbf{x_1}, \mathbf{x_2}$. Thus, once we know the product-moment correlation, we can calculate the angle between $\mathbf{x_1}$ and $\mathbf{x_2}$ in an orthogonal coordinate system by

$$ \theta_{12} = cos^{-1} r_{12}. $$

Thus, when the two variables are perfectly correlated, the two axes become one and the variates span the same space, but when the correlation is not perfect, the two axes of the two variables are separated by θ_{12}. This point is crucially important when we discuss the joint graphical display of quantification theory.

2.3 Standardized Versus Non-standardized PCA

Let us talk about standardization and non-standardization of input data for analysis. Consider the general case of n variables. Principal component analysis was developed for the multidimensional analysis of continuous variables. In practice, we typically consider two forms of principal component analysis, namely, principal component analysis of standardized variables and that of non-standardized or original variables.

This distinction is made primarily because there are two situations in which data are characterized:
(1) Data from social science research often lack the rational origin or unit (e.g., mathematics test scores, history test scores, personality scores, job evaluation, and food preference). To attain some comparability of those data, we use standardization of input data.
(2) Data with accurate measurement such as weight, height, or distance, which are generally seen in physical science data than in social science data, and the data (e.g., distance data) can be directly subjected to analysis.

These differences lead to two types of principal component analysis, one is to solve the eigenequation of the $n \times n$ correlation matrix **R** and the other of the $n \times n$ variance-covariance matrix **V**. These correspond to the two forms of linear combinations, one with standardized variables and the other with raw (original) variables. We will not derive principal component analysis for each of the two types here, but simply mention that principal component analysis (PCA), explained above, is equivalent to solving the following eigenequations:

- PCA of the Correlation Matrix **R**:

$$(\mathbf{R} - \lambda_k^2 \mathbf{I})\mathbf{y_k} = \mathbf{0}, \text{ or, } \mathbf{RY} = \mathbf{Y}\Lambda^2,$$

where λ_k^2 is the kth eigenvalue, **I** is the identity matrix, $\mathbf{y_k}$ is the corresponding eigenvector, **Y** is the matrix of eigenvectors, and Λ^2 is the diagonal matrix of eigenvalues in the main diagonals. It is typical that we impose the condition that $\mathbf{Y'Y} = \mathbf{I}$, that is, the vector $\mathbf{y_k}$ is standardized, namely, $\mathbf{y'_k y_k} = 1$. This means that $\mathbf{Y'RY} = \Lambda^2$.

- PCA of the Variance-Covariance Matrix **V**:

$$(\mathbf{V} - \lambda_k^2 \mathbf{I})\mathbf{y_k} = \mathbf{0}, \text{ or, } \mathbf{VY} = \mathbf{Y}\Lambda^2,$$

where λ_k^2, $\mathbf{y_k}$, $\mathbf{Y}$, and Λ^2 are, respectively, the quantities associated with corresponding terms of the variance-covariance matrix.

Caution!

One should be aware, however, that principal components of the same data set, associated with **V** and **R**, are in general very different. In other words, standardization can substantially alter the principal component structure of the data. Thus, the two outcomes associated with the same data can become totally incomparable by the simple transformation of the input data. Nishisato and Yamauchi (1974) used numerical examples to demonstrate how much the two PCA outcomes from the same data set might become utterly non-comparable once the data are standardized. We wonder which outcome, associated with **V** or **R**, can be regarded as the data structure we are looking for. The fact that one must ask such a question is amazing, isn't it? More surprising is the fact that we do not know how to overcome this simple-looking problem.

So, let us state again that the PCA results from **R** and **V** are in general not comparable, and that the onus of the decision on which one to analyse rests on the shoulders of the researchers. The problem of standardization or non-standardization looks like an innocent question to the researchers, but it poses a serious consequence for interpretations of the analytical results. One should keep this tricky problem in mind.

In leaving this problem of standardization associated with PCA, we should not overly be worried when we discuss quantification theory, for the situation is quite different from that of continuous data. We transform the categorical data in a specific way as we will see in the ensuing chapters. In other words, we will not have to deal with the standardizing problem as we saw in PCA.

We have just introduced orthogonal axes for the description of data. In the context of PCA, we have also learned that the attraction of the PCA procedure lies in the fact that it is the most efficient way of describing multidimensional data because of its typical extraction of components from the most important (component with the maximal variance) to the least important—in practice, we often look at only several important components.

2.4 Principal Versus Standard Coordinates

To simplify our discussion, let us consider that the variables are standardized. Then, let us indicate by **X** the matrix of standardized scores of N subjects on n tests, namely each test has the variance of 1 so that

$$\frac{1}{N}\mathbf{X}'\mathbf{X} = \mathbf{R} = \mathbf{Z}'\Lambda^2\mathbf{Z} = \mathbf{Y}'\mathbf{Y}$$

is the correlation matrix, where $\mathbf{Z}$ is called the matrix of *standard coordinates* which satisfies $\mathbf{ZZ}' = \mathbf{I}$ and $\mathbf{Y}$ is the matrix of *principal coordinates* such that $\mathbf{Y} = \mathbf{Z}\Lambda$ and $\Lambda = diag(\rho_j)$, where the elements of Λ are called singular values.

Let us look at the *geometry* of principal component analysis of the standardized variables. Since the data are standardized with respect to each of the n variables, we expect to observe the following:

- If the data are two-dimensional and we plot the variables in a two-dimensional graph, all the variables lie at the distance of 1 from the origin on the circle of radius 1.
- If the data are three-dimensional, all the variables lie on the surface of the three-dimensional sphere at the distance of 1 from the origin.
- No matter how many dimensions the data are scattered over, each data point lies at the distance of 1 from the origin, irrespective of its location in space.

The following two two-dimensional graphs show the plots of principal (regular circles) and standard (elongated circles) coordinates of two standardized variables.

This situation in which all standardized variables are located at the distance of 1 from the origin can be realized only when we plot the *principal coordinates* of variables. To repeat, please note that the data points positioned at the distance of 1 from the origin happen if and only if we use the principal coordinates, not the standard coordinates. In other words, *the data structure can be depicted only when we use principal coordinates*. This point is thoroughly discussed by Nishisato (1996).

Standard coordinates are introduced in the process of mathematical derivation of principal components, and they are standardized with respect to all the dimensions (components). Therefore, the visual appearance of the plot of data in terms of standard coordinates is totally different from our expectation. For example, consider the case in which the first eigenvalue is relatively large (namely, the concentration of data points on the first axis is comparatively great) and the second eigenvalue relatively small, the standard coordinates compensate for the distributions of the eigenvalues, leading to the situation that the two-dimensional plot of the data would show many data points close to the first axis while data points on the second axis are farther from the origin to compensate for the smaller number of variables with comparatively large loadings on the second component, as we see in Fig. 2.2. Imagine that even in this case, the plot of principal coordinates of the variables shows a perfect circle.

Thus, in multidimensional data analysis, we must *always use principal coordinates* to describe the data structure, and not standard coordinates. This is vitally important whenever we consider drawing graphs of variables in data analysis. So, let us remember:

- Principal coordinates are the ones that depict the structure of the data.
- Standard coordinates are not the ones to be used to describe the data structure.

This distinction is crucially important for the graphical display of data and can be used to judge against the use of a currently popular method of joint graphical display called *non-symmetric scaling*, as we will see in the next chapter.

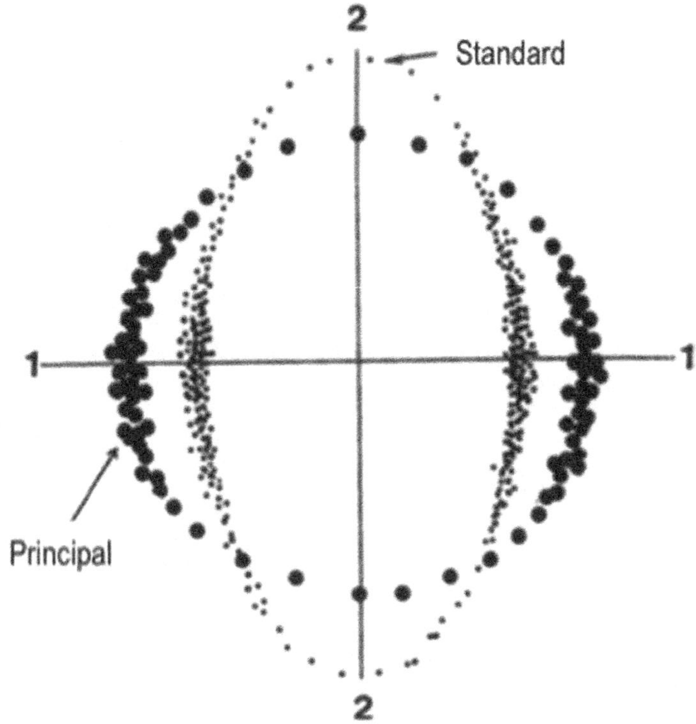

Fig. 2.2 Principal versus standard coordinates (from Nishisato 1996)

In terms of graphical display, there is one key difference between PCA and quantification theory. This difference comes from the fact that PCA is typically a uni-modal analysis, while quantification theory is bi-modal. Thus, the methods employ the following decompositions:

- PCA: the variance-covariance matrix is decomposed into $\mathbf{Y}'\mathbf{Y} = \mathbf{Z}'\Lambda^2\mathbf{Z}$ so that the matrix of principal coordinates is given by $\Lambda\mathbf{Z}$. This is the decomposition of $\mathbf{Z}'\Lambda^2\mathbf{Z}$, that is, $\mathbf{ZZ}' = \Lambda^2$.
- Quantification theory: The standardized frequency table is given by $\mathbf{Y}'\Lambda\mathbf{X}$, so that we either obtain $\mathbf{Y} = \Lambda\mathbf{Z}$ and $\mathbf{X}$, or $\mathbf{Y}$ and $\mathbf{X} = \Lambda\mathbf{Z}$. This is the decomposition of $\mathbf{Y}'\mathbf{X}\Lambda$.

Notice the difference in which the squared singular value appears in PCA, while the singular value is used in quantification theory. This distinction will later prove that it is the source of the perennial problem of joint graphical display in quantification theory.

In summary, we have seen that we can introduce an orthogonal coordinate system for a set of correlated variables, using principal axes. Operationally, we consider a linear combination of variables subject to the condition that the sum of squares of weights for the variables be 1. We determine the weights in such a way that the variance of the linear combination be a maximum. Once we derive such linear combinations of the variables, we obtain principal coordinates of the variables, which can be used for a multidimensional graph of data in the Euclidean space.

References

Hotelling, H. (1933). Analysis of complex of statistical variables into principal components. *Journal of Educational Psychology*, *24*, 417–441, 498–520.

Mardia, K. V., Kent, J. T., & Bibby, J. M. (1979). *Multivariate Analysis*. London: Academic Press.

Nishisato, S. and Yamauchi, H. (1974). Principal components of deviation scores and standard scores. *Japanese Psychological Research, 16*, 162–170.

Nishisato, S. (1996). Gleaning in the field of dual scaling. *Psychometirka, 61*, 559–599.

Pearson, K. (1901). On lines and planes of closest fit to systems of points in space. *Philosophical Magazines and Journal of Science, Series*, *6*(2), 559–572.

Rao, C. R. (1952). *Advanced Statistical Methods for Biometric Research*. New York: Wiley.

Takeuchi, K., Yanai, H., & Mukherjee, N. N. (1982). *The Foundations of Multivariate Analysis: A Unified Approach by Means of Projection Onto Linear Subspaces*. New Delhi: Wiley Eastern.

Chapter 3
Bi-modal Quantification and Graphs

To start with, we should admit that LS appears to make sense, but that it is not a general enough method for data analysis. Without any introduction, we can state that QT provides the most effective remedy for LS by providing optimal transformation of input data and using exhaustive analysis of information in data. This comparison between the two methods, LS and QT, will open our eyes to an extensive array of possible applications of QT to practical problems. At this moment, we should know that QT is a mathematically optimal means for extracting information from the data. But, how and why is it?

After we look at the basic aspects of QT, we will pay attention to a persistent problem associated with the use of QT, namely the joint graphical display of quantification results. This problem has plagued its sound developments for many years, and it is the main object of Part 1 to settle the matter once and for all by offering a solution to the problem. In order to attain this objective, we will gradually build our ladder step by step towards the solution at the apex. We will not leave even the slightest doubt about our solution through precise geometrical analysis of multidimensional space. This struggle towards the solution has been a personal history of at least one person, the author of Chaps. 1–5. It is hoped that the solution presented here will further push forward the frontier of QT research.

As may be clear from Chaps. 1 and 2, our focus is on bi-modal analysis of the contingency table, that is, equal handling of both rows and columns of the data matrix. Thus, this is a well-defined task, but we immediately wonder why it has taken so long to arrive at the current stage of development. Why have most researchers been satisfied with the compromised practice of joint graphical display? All the discussions in this chapter will provide a useful introduction to our task, that is, finding exact Euclidean coordinates for row variates and column variates of the contingency table in common space.

S. Nishisato et al., *Modern Quantification Theory*,
Behaviormetrics: Quantitative Approaches to Human Behavior 8,
https://doi.org/10.1007/978-981-16-2470-4_3

3.1 Likert Scale

Data analysis has been carried out in many research fields. When data are collected, the set of numbers often requires some forms of transformations before analysis, ranging from simple counting to logical analysis. Shortly after 1930, there was an epoch-making proposal.

In 1934, Likert completed his Ph.D. thesis in which he devised a convenient way to collect data, known as the Likert scale:

- Use integers as scores to ordered response categories such as 1 to "never", 2 to "sometimes", 3 to "usually", 4 to "often", and 5 to "always" to such questions as "How often do you wash your hands?" or "Do you sleep well?"

These scores are now referred to as Likert scores and the ordered set of integers is called the Likert scale. This invention made it easy to code responses to ordered response categories, and Likert scores have since been used in many such survey questionnaires such as in educational, medical, sociological, and psychological research. It was indeed a convenient way for coding *ordered* categorical data (i.e., it does not apply to non-ordered categories such as [male, female] and [urban, rural]).

3.1.1 Its Ubiquitous Misuse

Remember that the Likert scale was proposed at the time when the researchers were mainly concerned with constructing *unidimensional* scales of such attributes as anxiety, personality, and political attitudes. In fact, Guttman (1950) also developed a method of unidimensional scaling, called scalogram analysis, applied typically to binary data. Here, again the data were collected to construct unidimensional scales of psychological attributes (e.g., attitudes towards social justice and religious beliefs). Those days, we were interested in collecting such questions that the responses to them conform to a test of a unidimensional attribute (e.g., Guttman's coefficient of reproducibility (Guttman 1950) and Cronbach's reliability coefficient (Cronbach 1951)). Our main task then was to collect questions that satisfy a unidimensional criterion (e.g., Guttman's high reproducibility and Cronbach's high reliability). Then the total scores (if the question is binary, the sum of "yes" responses; if the question is of the rating form, then the sum of the Likert scores) were considered as reasonable scores for interpretations.

Those were typical studies in the early days, but the research interest was gradually extended from unidimensional traits to multidimensional traits as we can infer from such a new method as factor analysis (e.g., Thurstone 1947; Harman 1960). In spite of this dramatic transition in the research orientation, the old-fashioned Likert scale remained popular as a scoring method. For multidimensional analysis, Likert scores such as 1, 2, 3 are only appropriate for the purpose of *coding* of the input data, but not for calculating statistics. This is a very important point and thus should be emphasized. Unfortunately, however, this point was ignored by most researchers.

To repeat, the Likert scale is nowadays useful only as a coding method, and it no longer serves as a scoring method. Unless we are looking for unidimensional attributes, Likert scores of 1, 2, 3, 4, and 5 for five ordered categories can no longer offer appropriate data for the derivation of meaningful statistics.

3.1.2 Validity Check

We must examine if Likert scores can be meaningfully subjected to quantitative analysis. This check is easy and must be done prior to analysis. Currently, most researchers use Likert scores simply because they look reasonable, and do not worry about the consequences. We would like to present three numerical examples of Likert scores, one appropriate and the other two not appropriate. Let us start with a case in which Likert scores are appropriate.

Example 1: Likert scores are appropriate

This example is from Nishisato (1980). Suppose that from a survey on the use of sleeping pills, 28 subjects were randomly sampled from each of the following five groups:

(1) those strongly against the use of sleeping pills;
(2) those moderately against;
(3) those neutral;
(4) those moderately for;
(5) those strongly for.

Those subjects were then asked the question: "Do you have nightmares in sleep?" The answer to this question is made by choosing one of the following response alternatives: (1) never, (2) rarely, (3) sometimes, (4) often, and (5) always. The 5×5 contingency table of their responses is as shown in Table 3.1.

Table 3.1 Contingency table of sleeping pills and nightmares

	Nightmare						
	Never	Rarely	Sometimes	Often	Always	Total	Score
Strongly against	15	8	3	2	0	28	−2
Against	5	17	4	0	2	28	−1
Neutral	6	13	4	3	2	28	0
For	0	7	7	5	9	28	1
Strongly for	1	2	6	3	16	28	2
Total	27	47	24	13	29	140	
Score	−2	−1	0	1	2		

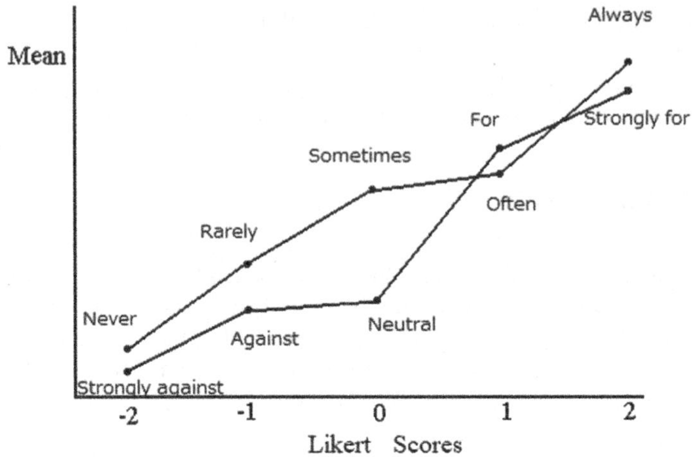

Fig. 3.1 Means against Likert scores

Given those Likert scores of [−2,−1,0,1,2] in the table, let us calculate the means of the rows and those of the columns. For example, the mean value of "strongly against" can be calculated from the response frequencies (15, 8, 3, 2, 0) and the corresponding Likert scores as

$$\frac{15 \times (-2) + 8 \times (-1) + 3 \times 0 + 2 \times 1 + 0 \times 2}{28} = -1.3.$$

Similarly, the mean value of "never" is calculated from the five frequencies (15, 5, 6, 0, 1) in column "never" and the corresponding Likert scores as

$$\frac{15 \times (-2) + 5 \times (-1) + 6 \times 0 + 0 \times 1 + 1 \times 2}{27} = -1.2.$$

In this way, we can obtain five row means as [−1.3, −0.8, −0.6, 0.6, 1.1] and column means [−1.2, −0.5, 0.4, 0.5, 1.3]. The appropriateness of Likert scores can be checked by plotting these means against Likert scores [−2, −1, 0, 1, 2] as shown in Fig. 3.1. Notice that the two line plots are relatively linear with positive slopes. In this case, we can state that Likert scores are appropriate for this data set. Our conclusion then is that the opinion about taking sleeping pills is *linearly related* to the frequency of nightmares. Alternatively, we can state that Likert scores are appropriate when the two variables are linearly related.

We will later see that quantification theory "adjusts" those Likert scores further in such a way that the plots of the means against the modified Likert scores are perfectly linear—we will find out how important this simple transformation means for data analysis.

Table 3.2 Contingency table of weight lifting and age

Maximum weight	W1	W2	W3	W4	Sum
A1	16	9	3	0	28
A2	1	16	25	6	48
A3	0	12	18	26	56
A4	15	20	3	0	38
Sum	32	57	49	32	170

Notes: A1 = Up to 15 years old; A2 = 16–40; A3 = 41–65; A4 = Over 65
W1 = Up to 10 kg; W2 = 16–40 kg; W3 = 41–65 kg; W4 = Over 65 kg

Example 2: Likert scores are not appropriate

Suppose 170 people were asked about their ages (Q.1) and how much weight they can lift with a single arm (Q.2). The following artificial data are created for this demonstration. Note that the two sets of the response categories are all ordered:
Age: younger than 15, 16–40, 41–65, older than 65;
Weight lifting: up to 10 kg, 16–40 kg, 41–65 kg, over 65 kg.

For this type of data, most researchers use Likert scores, and assign, for example, 1 to "up to 15 years old", 2 to the age 16–40, 3 to the age 41–65, and 4 to the age older than 65. Similarly, we assign 1 to the weight up to 10 kg, 2 to the weight 16–40 kg, 3 to the weight 41–65 kg, and 4 to the weight over 65 kg. These are Likert scores and they look reasonable. In fact, there is nothing wrong with this assignment *so long as we treat them as codes*. The problem may arise if the age and the weight lifting are not linearly related. As in the previous example, let us use the Likert scores for calculating the average scores of the row and column categories (Table 3.2).

$$m_{A1} = \frac{16\times 1 + 9\times 2 + 3\times 3 + 4\times 0}{28} = 1.5, m_{A2} = 2.8, m_{A3} = 3.4, m_{A4} = 1.7.$$

Similarly, we can calculate the average column scores as

$$m_{W1} = \frac{16\times 1 + 1\times 2 + 0\times 3 + 15\times 4}{32} = 2.4, m_{W2} = 2.8, m_{W3} = 2.4, m_{W4} = 2.8.$$

When we plot these means on the vertical axis and the Likert scores for the two sets of categories on the horizontal axis, we obtain Fig. 3.2. Recall that we wanted to capture the tendency through data analysis that the strength of people is weak when one is very young, and then increases as one gets older, but declines as one gets further older. But, the analysis through Likert scores will never capture that nonlinear relation between the strength and the age. All we get from a typical analysis is there does not seem to be any relation between the age and the strength, which is absurd.

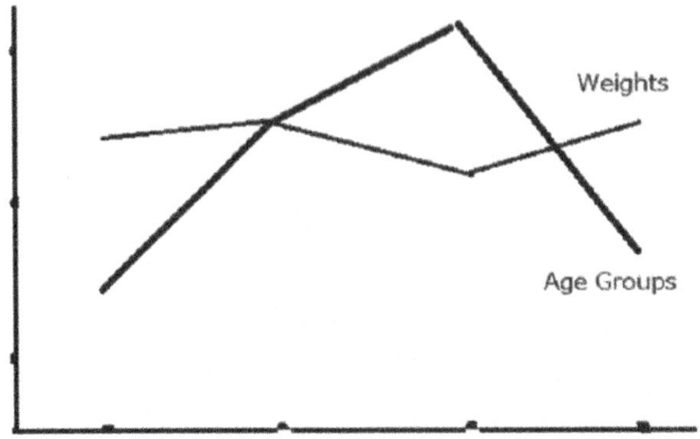

Fig. 3.2 Relation between Likert scores and the means

The two connected lines for the rows and the columns are far from the straight line. This means that Likert scores are neither appropriate nor useful for capturing the information in this data set. Why not? Because Likert scores are simple substitutes of quantities for linearly related variables, and the current variables (age and strength) are not linearly related to each other. By simple inspection of the data, we can easily tell that the arm strength first increases as one gets older and then decreases when one gets further older. This is definitely an example of a nonlinear relation between the two variables, strength and age. Therefore we cannot use Likert scores to grasp the information in the data.

Thus, the fact that the response categories are ordered does not mean that the relations between variables are linear.

Example 3: Likert scores are not appropriate

Suppose we ask people about their preference for coffee with various water temperatures. Some people would like ice coffee (cold) and also reasonably hot coffee (hot), but most people would not like frozen coffee (extremely cold), lukewarm coffee (mildly hot), or boiling-hot coffee (extremely hot). So, our expectation is that preference for coffee is a nonlinear function of the temperature of the water. Suppose we offer subjects cups of coffee at different temperatures of water and ask them to rate how much they like each coffee. The 5×5 contingency table of their responses is as shown in Table 3.3, where the temperature in Likert scale is -2 = frozen, -1 = cold, 0 = lukewarm, 1 = hot, 2 = boiling hot, and the preference in Likert scale is -2 = worst, -1 = terrible, 0 = not good, 1 = good, and 2 = best. Given those Likert scores in the table, let us calculate the means of the rows and those of the columns as before. We obtain five row means as [−2.0, 1.5, 1.1, 1.7, −1.0] and five column means as [0.0, 0.1, −0.3, −0.1, −0.3]. If we look at the preference values (means) of coffee, the results show that hot coffee is most preferred, then cold (ice) coffee, then lukewarm coffee, then boiling hot coffee, and the last is frozen coffee, showing that the preference is a nonlinear function of the water temperature. The plot of these

Table 3.3 Preference of coffee at different water temperatures

	Preference						
	Worst	Terrible	Not Good	Good	Best	Total	Score
Frozen	29	1	0	0	0	30	−2
Cold	0	0	2	10	18	30	−1
Lukewarm	8	18	4	0	0	30	0
Hot	0	0	0	9	21	30	1
Boiling	28	2	0	0	0	30	2
Total	60	22	6	19	39	150	
Score	−2	−1	0	1	2		

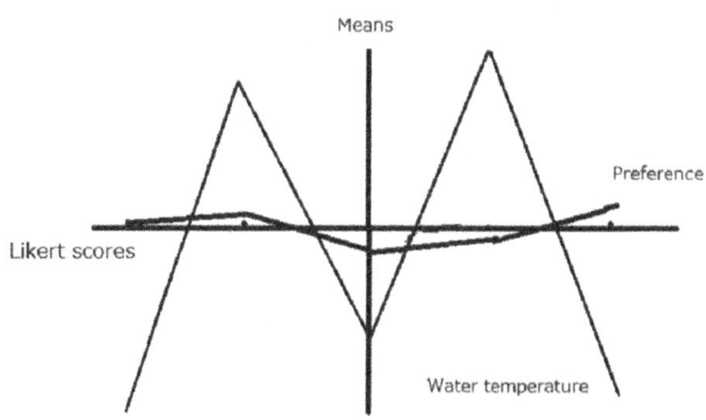

Fig. 3.3 Means against likert scores

means against Likert scores [−2, −1, 0, 1, 2] is as shown in Fig. 3.3. The two lines are clearly not linear, indicating that the Likert scores are not appropriate for this example. As we see in this example, Likert scores would not allow us to see that most people like hot coffee, some prefer ice coffee to hot coffee, but most people would avoid lukewarm coffee. This kind of common-sense expectation can never be expressed by data with Likert scores. In other words, the preference for coffee is not a linear function of the temperature of water as Likert scores imply.

In summary, we can conclude that Likert scores are appropriate only when the phenomena we are investigating are linearly related. Surprisingly in most survey data analyses, the appropriateness of Likert scores is not even checked and conclusions on analysis are frequently made based on inappropriate use of Likert scores. We should be aware that so many nonlinear relations are involved in most data sets we analyse. Thus, the Likert scale yields appropriate scores when the phenomenon we are dealing with is linear relations. Otherwise, Likert scores should be used only as *codes* that require further transformation.

One lesson we learn from the above observations is that we must look for a method which can capture the underlying information whether the data involve linear or nonlinear relations. Where can we find such a flexible method?

3.2 Quantification Theory

One of them is nothing but quantification theory (QT) itself, for it quantifies information in data so as to best capture the relations, linear or nonlinear, between variables. In fact, QT can deal with any kinds of relations embedded in data. This is a stark contrast to a widely used analysis of Likert scores (LS), of which proper applications are so limited that it is almost useless as a scoring method: its use is limited only as a coding method for ordered sets of response categories and not as a scoring method.

Another advantage of QT over LS is that QT can handle not only **ordered categories** such as (never, sometimes, always) and (strongly disagree, moderately disagree, neutral, moderately agree, strongly agree), but also **non-ordered categories** such as (female, male), (rural, urban), (coffee, tea, juice), and (right-handed, left-handed, ambidextrous). [Note: Nishisato 1980, 1994 deals, under the umbrella of QT, with such categorical data as rank-order data, paired comparison data, sorting data, successive categories data, and multi-way data]. Since LS deals only with linear relations of data collected by ordered categories, we wonder why LS is still widely used. In contrast, QT can handle not only a variety of categorical data but also any functional relations between variables.

3.2.1 Quantification by Reciprocal Averaging

Let us now introduce a simple procedure of QT, called the method of reciprocal averages (Richardson and Kuder 1933; Horst 1935; Mosier 1946). This procedure was based on the application of a convergent sequence of numbers, proposed when computers were not as generally available as today. As we will see, it gives us an intuitive grasp of what quantification means, and as such it will give us a good introduction to QT. The method consists of the following steps: Given a contingency table as data,

1. Assign arbitrary scores to the rows (avoid identical scores).
2. Calculate the averages of the columns, weighted by these row scores.
3. Using the new column averages as scores, calculate the new row averages.
4. Using the new row averages as scores, calculate the new column averages.
5. When the successive row and column averages converge, respectively, to the stable row and column values, then these final values are *optimal* scores for rows and columns.

In this iterative process, it is typical that row scores and column scores are normed to a certain constant so that the process will not diverge. See the details in Nishisato (1984, 1986, 1994) and Nishisato (1982, 1994, 2007), where they implemented the reciprocal averaging process with the successive standardization so that the convergence of the process can be observed by the identical numerical values to those from the previous iteration. We will soon see that this early invention has an important

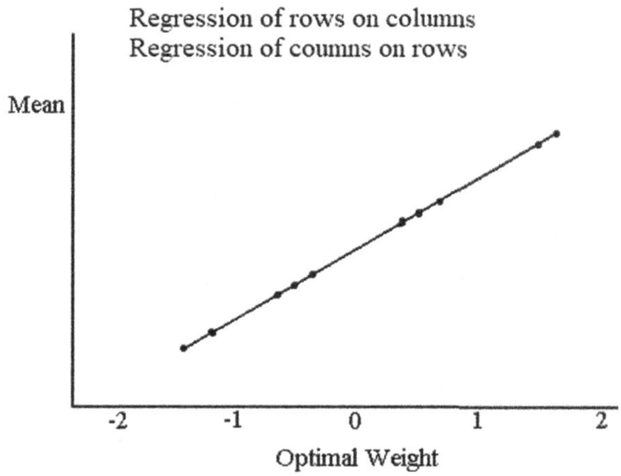

Fig. 3.4 Means against optimal scores

implication for the joint graphical display. Before we find it out, let us indulge for a moment in the numerical illustration of the procedure.

There is a mathematical proof that the reciprocal averaging process eventually converges to two maximal linear relations between the two variables (Nishisato 1994): at this stage, plot the means of rows against column scores and the means of columns against row scores, which we will find to coincide into a single straight line and are *perfectly linear*. Once the convergence is obtained, we can declare that we have obtained the optimal solution. Rather than showing this iterative process here, we will present only the results when the convergence is attained, using the example of the sleeping pills and nightmares (Table 3.1 and Fig. 3.1). See Fig. 3.4 where the two lines (one for sleeping pills and the other for nightmares) coincide with the slope equal to the maximum correlation between rows and columns. Note that the optimal weights of rows are [−1.20, −0.64, −0.49, 0.87, 1.47] for (strongly against, against, neutral, for, strongly for) and those of columns are [−1.30, −0.59, 0.43, 0.58, 1.55] for (never, rarely, sometimes, often, always], respectively, and the singular value is 0.65.

Notice that the slope of the graph is equal to the singular value, which is 0.65 and it is also equal to the maximized product-moment correlation. When we used Likert scores, the correlation was 0.63. The fact that Likert scores yielded the value of correlation coefficient very close to the maximal value of 0.65 indicates that Likert scores are appropriate for this example. One should be warned at this stage about the fact that the two lines merging into a single line may give us the impression that both rows and columns span the same space, but this is not the case! It simply shows that both rows and columns are symmetrically scaled (quantified).

Note that even for those cases in which Likert scores were totally inappropriate (see the examples of the body strength and age, the preference of coffee in terms of the water temperature), we obtain perfect straight lines when we plot the means against the optimal category weights. This means that our quantification rationale is to perform a nonlinear transformation of quantities assigned to rows and columns in such a way that the relation between rows and columns becomes perfectly linear.

These scores on the horizontal axis are called *optimal scores* for rows and columns of the contingency table. Optimal scores have the following characteristics:

1. The slope in the graph is equal to the largest *singular value*, ρ, of the contingency table.
2. The slope is equal to the *maximal correlation* between the rows and the columns of the contingency table.
3. The arc cosine of the slope, $cos^{-1}\rho$, is the *angle*, θ, between the axis for rows and the axis for columns in multidimensional space. In other words, rows and columns do not in general span the same space! Thus, each QT component is two-dimensional, typically with oblique axes, one for rows and the other for columns.
4. The slope is the *projection operator* of the rows onto the columns and vice versa.
5. This mutual projection of one over the other means that quantification is carried out symmetrically over the rows and the columns.

When the process converges to the stable values, the corresponding sets of row weights and column weights are called the first component. Then, we calculate the residuals defined as the differences between the original data minus data accounted for by the first component, which are then subjected to the reciprocal averaging operation, yielding the second component. The same reciprocal averaging process is applied to the residual frequencies successively until no more information is left to analyse. From the nxm contingency table, we can extract p components, where $p =$ the smaller value of $(n - 1)$ and $(m - 1)$. Since the rows and the columns are generally not perfectly correlated in individual dimensions (components), we need 2K-dimensional space for K-component data.

Out of those points listed above, the most relevant to our discussion is the third point that the row axis and the column axis are separated by a non-zero angle unless the singular value is 1. This is a crucial point for our discussion on joint graphical display.

• **Notes:** The method of reciprocal averages (MRA) was later generalized in two ways:
(1) **Piecewise method of reciprocal averages: PMRA** (Nishisato and Sheu 1980). This is an extension of MRA for the contingency table to MRA of multiple-choice data or multi-way contingency tables.
(2) **Method of reciprocal medians: MRM** (Nishisato 1984). This method replaces averages with medians for robust quantification.

3.2.2 *Simultaneous Linear Regressions*

Rather than pursuing the successive transformations of the reciprocal averaging process to the linear relations, Hirschfeld (1935) posed the question if we can transform the contingency table in such a way that the regression of rows onto columns and the regression of columns onto rows be simultaneously linear. In other words, Hirschfeld derived a mathematical equation for simultaneous linear regressions. His solution is exactly identical to the results we obtain from the reciprocal averaging when the convergence is attained. We can therefore say that Hirschfeld provided an algebraic solution to the quantification problem without any iterative process. With his formulation, we can now look at the well-known bi-linear expansion of the frequency table.

3.2.2.1 Horst and Hirschfeld

Horst used to tell Nishisato about the early days when he and Richardson worked at the Proctor and Gamble Corporation and used their *method of reciprocal averages* for data analysis. H.O. Hartley was a famous statistician, and Nishisato's greatest regret was that he did not know that H.O. Hartley was the same person as H.O. Hirschfeld of *simultaneous linear regressions*! Nishisato still remembers Hartley's vibrant personality as demonstrated in his starting words at a conference held at the University of Wisconsin: "Can you hear me? I will not ask you if you can see me" which caused a big laughter among the audience. It was a large classroom with no raised platform nor the microphone—he was very loud, but people in the back seats could not see even the top of his head. Both Horst and Hartley (Hirschfeld) were friendly, modest, and great scholars.

3.3 Bi-linear Decomposition

Let us look at an $m \times n$ contingency table F of frequencies with typical element f_{ij} for row i and column j. Our objective is to assign weights for the rows and the columns of the contingency table in such a way that the row weights and the column weights together account maximally for the frequencies in the table. This is basically the same criterion used in PCA, in which we maximize the variance of the composite scores, thus minimizing the differences between observed scores and derived scores. The major difference between PCA and QT (quantification theory) is that the former is typically uni-modal analysis, while the latter is bi-modal analysis. In other words, PCA is devoted to the multidimensional structure of only one of the two sets of variables, that is, either the rows or the columns of the data matrix, while QT places the same importance on both row and column variables. From the viewpoint of graphical display, the major difference between PCA and QT is whether we wish to find coordinates of only row variables (or column variables), or both row and column variables—remember that rows and columns of the contingency table are typically correlated, but not perfectly, necessitating additional space to accommodate correlated variables. Whether the analysis is uni-modal or bi-modal thus makes a significant difference from the graphical point of view, and this is perhaps the greatest difference between PCA and QT.

With respect to this aspect, QT is *bi-modal principal component analysis*. Although Torgerson (1958) called QT a PCA of categorical data (Torgerson 1958), it should more precisely be called bi-modal PCA of categorical data.

The bi-linear expansion of the contingency table with typical element f_{ij} is the same as singular value decomposition of the contingency table and can be expressed as

$$f_{ij} = \frac{f_{i.}f_{.j}}{f_t}(1 + \rho_1 y_{1i}x_{1j} + \rho_2 y_{2i}x_{2j} + \rho_3 y_{3i}x_{3j} \cdots + \rho_K y_{Ki}x_{Kj}).$$

Notice the symmetric (dual) decomposition, which is the essence of the bi-modal aspects of our analysis. In terms of matrix notation, we have the following:

$$\mathbf{F} = \left(\frac{1}{f_t}\right)\mathbf{D_r}\mathbf{Y}\Lambda\mathbf{X}'\mathbf{D_c}$$

where $\mathbf{F}$ is the $m \times n$ contingency table, f_t is the sum of the elements of the contingency table, $\mathbf{D_r}$ is the $m \times m$ diagonal matrix of row totals of the contingency table, $\mathbf{Y}$ is the $m \times K$ basis matrix such that $\mathbf{Y}'\mathbf{Y} = \mathbf{I}$, $\mathbf{D_c}$ is the $n \times n$ diagonal matrix of column totals of $\mathbf{F}$, $\mathbf{X}$ is the $n \times K$ basis matrix such that $\mathbf{X}'\mathbf{X} = \mathbf{I}$, and Λ is the $K \times K$ diagonal matrix of singular values.

Let us note the following outcomes of the expansion:

- The matrix of principal coordinates of rows, say $\mathbf{Y}^*$, is given by

$$\mathbf{Y}^* = \mathbf{Y}\Lambda.$$

- The matrix of principal coordinates of columns, say $\mathbf{X}^*$, is given by

$$\mathbf{X}^* = \mathbf{X}\Lambda.$$

- The matrix of standard coordinates of rows is given by $\mathbf{Y}$.
- The matrix of standard coordinates of columns is given by $\mathbf{X}$.

At this stage, note that the key component of the contingency table $\mathbf{Y}\Lambda\mathbf{X}'$ can be expressed as

$$\mathbf{Y}\Lambda\mathbf{X}' = (\mathbf{Y}\Lambda)\mathbf{X}' = \mathbf{Y}^*\mathbf{X}' = \mathbf{Y}'(\Lambda)\mathbf{X} = \mathbf{Y}'\mathbf{X}^*.$$

In other words, the contingency table can be decomposed into the product of [the *principal* coordinates of the rows and *standard* coordinates of the columns], or [*standard* coordinates of the rows and *principal* coordinates of the columns]. Please note here that the data structure of QT cannot be expressed as the product of principal coordinates of *both rows and columns*, the condition under which the rows and the columns can be accommodated in the common space. This will become an important aspect of QT when we discuss joint graphical display later.

The above decomposition is well accepted and widely used for QT of the contingency table. In most algorithms currently in use, the first component maximally accounts for the joint frequency, the second component explaining maximally the residual information in data, and so on in the descending order of importance, where the importance is measured by the eigenvalue ρ^2, which is an indicator of the amount of information accounted for by component k. With the complete expansion given by the above formula, we often say that QT is a method to provide exhaustive orthogonal analysis of information in data.

3.3.1 Key Statistic: Singular Values

Since mathematics of QT has been discussed in many publications (see references cited in Chap. 1), we will only note the role of singular values in quantification theory. Let us list the properties of singular value ρ:

1. Singular value ρ_k is the correlation between rows and columns of component k. This is conditionally the maximal product-moment correlation for component k.
2. ρ_k is the projection operator for rows onto columns and columns onto rows for component k.
3. Therefore, the singular value can be used to calculate the discrepancy angle in degrees (θ) between the row space (vector) and the column space (vector) for component k (Nishisato and Clavel 2003),

$$\text{Discrepancy Angle:} \quad \theta_k = cos^{-1}\rho_k.$$

4. The square of ρ_k is the eigenvalue, which is the variance of the quantified variates of component k, namely the amount of information accounted for by component k.
5. The eigenvalue is also the correlation ratio (the between-rows sum of squares over the total sum of squares, or the between-column sum of squares over the total sum of squares).

Thus, the singular values and the squared singular values (i.e., eigenvalues) play crucially important roles as *sources of information* in data decomposition and as structural parameters of the data.

Table 3.4 Age and body strength: three components

Components	1	2	3
Eigenvalues	0.51	0.10	0.02
Singular values	0.72	0.32	0.15
θ (degree)	44°	71°	81°

3.4 Bi-modal Quantification and Space

The foregoing discussion indicates that unless the regression coefficient (singular value) is 1, the space for rows and the space for columns are separated by the arc cosine of the singular value, that is, $\theta = cos^{-1}\rho$. This means that we typically need two axes to represent a single component, one for row variables and the other for column variables. This has a direct implication for joint graphical display of row and column variates. Namely, we need a two-dimensional graph for each component!

Knowing that the two axes, one for the rows and the other for the columns, are separated by the angle mentioned above, how can we create a two-dimensional graph for each component?

Without any discussion on the quantification procedure and its outputs, let us look at the above example of the relation between age and the weight one can lift with a single arm, which yields three components with the decomposition information as shown in Table 3.4. The relevant information for our discussion is the discrepancy angle between row space and column space of each component. Since three components all show non-zero angles between row space and column space, we can conclude that this data set needs six-dimensional space to accommodate rows and columns in common space.

Generally speaking, the contingency table requires doubled multidimensional space for its graphical display. This is a stark contrast to the traditional belief that the number of non-zero singular values or the eigenvalues indicates the dimensionality of space. Remember this: for bi-modal quantification, the graphical display requires doubled number of dimensions.

3.5 Step-by-Step Numerical Illustrations

To familiarize ourselves with details of quantification analysis and space, let us look at a simple numerical example and follow through the entire spectrum of analysis. We hope that this section will provide us with a sufficient background for quantification analysis and our main topic, joint graphical display.

3.5.1 Basic Quantification Analysis

Recall that our ultimate goal is to arrive at the proper rationale behind joint graphical display. To convince the readers on the theory of doubled space, we will use numerical illustrations. Let us borrow the data from Garmize and Rychlak (1964). These authors investigated whether subjects' perceptions of Rorschach inkblots could be influenced by the moods of the subjects. The investigators devised a way in which different moods could be introduced into subjects' minds, and then studied how subjects' moods would influence the perceptions of Rorschach inkblots. The original data set included 16 Rorschach symbols and 6 moods. In analysing the 16×6 table of response frequencies, however, Nishisato (1994) noted that there were very few responses of Rorschach symbols of Bear, Boot(s), Bridge, Hair, and Island. Thus, he decided to delete those five symbols from his analysis in order to avoid possible outlier effects in quantification (Nishisato 1984, 1987, 1991). We follow his lead and will use the reduced data set of 11×6 table in Chaps. 3, 5, and 6. The data are summarized in the contingency table of joint occurrences of Rorschach inkblots and moods as shown in Table 3.5.

Using this data set, the quantification problem can be stated in a number of ways:

- We wish to derive scores for the 11 Rorschach responses and scores for the 6 moods in such a way that these scores will maximally account for the joint frequencies in Table 3.5 in the least-squares sense;
- We wish to determine these two sets of scores in such a way that when we apply them to the data in Table 3.5, the correlation between the rows and the columns of the table attains the maximal value;

Table 3.5 Rorschach data and induced moods (Garmize and Rychlak 1964)

	Induced	Moods				
Rorschach Responses	Fear	Anger	Depression	Love	Ambition	Security
Bat	33	10	18	1	2	6
Blood	10	5	2	1	0	0
Butterfly	0	2	1	26	5	18
Cave	7	0	13	1	4	2
Clouds	2	9	30	4	1	6
Fire	5	9	1	2	1	1
Fur	0	3	4	5	5	21
Mask	3	2	6	2	2	3
Mountains	2	1	4	1	18	2
Rocks	0	4	2	1	2	2
Smoke	1	6	1	0	1	0

Notes: Rorschach symbols Bear, Boot(s), Bridge, Hair, and Island were dropped from the original data set, due to small frequencies

Table 3.6 Summary statistic

Components	1	2	3	4	5
Eigenvalues	0.46	0.25	0.17	0.13	0.07
Singular values	0.68	0.50	0.41	0.36	0.27
δ	43%	23%	16%	12%	7%
$\Sigma\delta$	43%	66%	82%	93%	100%
θ (degree)	47°	60°	66°	69°	74°

$\chi^2 = 370.85$, df $= 50$, significant at $p = 0.05$

- We wish to determine the scores for the 6 rows in such a way that the variance of the 11 response variables be a maximum, relative to the total variance in the table;
- Alternatively, we wish to determine the 11 scores for the responses in such a way that the variance of the 6 moods be a maximum, relative to the total variance in the table;
- We wish to determine the two sets of scores in such a way that the regression of the 11 responses on the 6 moods and the regression of the 6 moods on the 11 responses be simultaneously linear;
- We wish to determine the scores for the 11 responses in such a way that the scored data attain the maximal value of Cronbach's generalized reliability coefficient (Cronbach 1951), as reported by (Lord 1958) and illustrated by Nishisato (1980);
- We wish to determine the scores for the 6 responses in such a way that the scored data attain the maximal value of Cronbach's generalized reliability coefficient.

These different optimization criteria, however, are known to yield the same *optimal values* for the 11 responses and the 6 moods. The total number of possible components from the contingency table is equal to "the smaller number of [the number of rows and the number of columns] minus 1". This "minus 1" corresponds to the so-called *trivial component*, which is associated with the case in which rows and columns are statistically independent, namely the case which we are not interested in, thus discarded as *trivial component*.

In the current example, the total number of components is 5. No matter which criterion for optimization (quantification) we may adopt, we will arrive at the following summary statistics (Table 3.6). The outputs are from Dual3 (Nishisato 1986). Unfortunately, the program package is no longer available for general distribution since the year 2000, but the copy of the companion book (Nishisato 1994) can be downloaded free from the LinkedIn site.

Together with the above results, we obtain coordinates of rows and columns in two forms, standard coordinates and principal coordinates, as shown in Table 3.7. Note that the standard coordinates are normed to the same constant for all components, and therefore they are not the quantities that can be used to describe the structure of the data. In contrast, principal coordinates are the quantities for interpreting the data structure.

Let us now look at the results step by step.

Table 3.7 Standard coordinates and principal coordinates

	Component	1	2	3	4	5
Standard	Bat	−1.03	−0.33	0.37	−0.95	−0.31
	Blood	−1.27	−0.71	1.45	−0.49	0.24
	Butterfly	1.71	−0.88	0.42	−0.43	1.18
	Cave	−0.55	0.60	−1.07	−0.96	0.42
	Clouds	−0.34	−0.15	−1.81	0.83	0.48
	Fire	−0.62	−0.60	1.52	1.66	0.25
	Fur	1.15	−0.15	−0.20	0.03	−2.50
	Mask	−0.08	0.08	−0.51	−0.12	0.09
	Mountains	0.50	3.08	0.76	−0.11	0.41
	Rocks	0.17	0.29	0.32	1.96	−0.29
	Smoke	−0.83	−0.10	1.38	3.48	0.02
	Fear	−1.27	−0.34	0.95	−1.34	−0.10
	Anger	−0.65	−0.47	0.85	2.08	−0.07
	Depression	−0.57	0.17	−1.64	0.06	0.36
	Love	1.51	−1.01	0.36	−0.32	1.80
	Ambition	0.62	2.53	0.70	0.00	0.20
	Security	1.11	−0.47	−0.21	−0.21	−1.75
Principal	Bat	−0.70	−0.16	0.15	−0.34	−0.08
	Blood	−0.87	−0.35	0.60	−0.18	0.06
	Butterfly	1.17	−0.44	0.17	−0.15	0.32
	Cave	−0.37	0.30	−0.44	−0.34	0.11
	Clouds	−0.23	−0.08	−0.75	0.30	0.13
	Fire	−0.42	−0.30	0.63	0.59	0.07
	Fur	0.78	−0.08	−0.08	0.01	−0.67
	Mask	−0.05	0.04	−0.21	−0.04	0.02
	Mountains	0.34	1.54	0.31	−0.04	0.11
	Rocks	0.11	0.15	0.13	0.70	−0.08
	Smoke	−0.57	−0.05	0.57	1.25	0.00
	Fear	−0.87	−0.17	0.39	−0.48	−0.03
	Anger	−0.44	−0.23	0.35	0.75	−0.02
	Depression	−0.39	0.09	−0.68	0.02	0.10
	Love	1.03	−0.51	0.15	−0.12	0.48
	Ambition	0.42	1.27	0.29	0.00	0.05
	Security	0.76	−0.23	−0.09	−0.08	−0.47

- **Chi-square:** When the chi-squared statistic for the contingency table is significant, it indicates that a non-random association between rows and columns exists. Recall that in bi-modal analysis, the total row-column relation is the primary object of our interest. Thus, this statistic is important for deciding whether it is worth decomposing information in data into orthogonal components. Once the significant total chi-square is obtained, it can be decomposed into components, in the current example, the significant total chi-square into five orthogonal components. As for the decomposition of the total chi-square into chi-squares of individual components, there are some arguments for and against, and the readers are referred to such papers as Bartlett (1947), Rao (1952), Williams (1952), Kendall (1957), Bock (1960), Lancaster (1953), and Kalantari et al. (1993).
- **Eigenvalues:** The eigenvalue of each component can be regarded as the amount of information, and mathematically it is the correlation ratio or the relative variance of the component. The corresponding singular values are square roots of the eigenvalues, and their meanings are as mentioned earlier. A direct interpretation of the eigenvalues may be difficult, and we typically calculate the percentage of information accounted for by each component as mentioned next.
- **Percentage of Information Accounted for:** The delta δ is the relative percentage of the variance accounted for by each component, and the cumulative delta indicates how much the total variance is accounted for by the components so far extracted. In our example, the first component accounts for 43 per cent of the total variance, and five components account for 100 per cent of information in data.
- **Data Reconstruction**
  Once we calculate the total information accounted for by percentage, however, it can be misleading, for some data contain more information, as reflected in the value of the eigenvalue, than others, but this is not reflected on δ. As a way to show the importance of each component, Nishisato (1980) (see also Nishisato and Nishisato 1994) provided the *kth order approximation* to the input data. Recall that the contingency table's typical element can be decomposed as follows:

$$f_{ij} = \frac{f_{i.}f_{.j}}{f_t}(1 + \rho_1 y_{1i}x_{1j} + \rho_2 y_{2i}x_{2j} + \rho_3 y_{3i}x_{3j} + \cdots + \rho_K y_{Ki}x_{Kj}).$$

  In this decomposition formula, we use the following definition:

  - The order-0 approximation $= \frac{f_{i.}f_{.j}}{f_t}$. This is the case in which rows and columns are statistically independent;
  - The order-1 approximation $= \frac{f_{i.}f_{.j}}{f_t}(1 + \rho_1 y_{1i}x_{1j})$;
  - The order-2 approximation $= \frac{f_{i.}f_{.j}}{f_t}(1 + \rho_1 y_{1i}x_{1j} + \rho_2 y_{2i}x_{2j})$;
  - The order-3 approximation $= \frac{f_{i.}f_{.j}}{f_t}(1 + \rho_1 y_{1i}x_{1j} + \rho_2 y_{2i}x_{2j} + \rho_3 y_{3i}x_{3j})$;

  and so on.
  This approximation shows what if we want to approximate the data under the assumption that the rows and the columns are statistically independent (order-0 approximation), how much the prediction will improve if the first component is

added to it (order-1 approximation), if the second component also is added to it (order-2 approximation), and so on. The order 1- to -4 approximations are listed in Table 3.8. In the current example, the order-5 approximation will reproduce the input contingency table perfectly (Table 3.5). This means that the input data can be decomposed into those components of order-k approximations. By looking at the progression of the order-k approximation, we can get some ideas as to the relative importance of individual components.

3.6 Our Focal Points

So far we have followed an ordinary procedure of quantification of contingency tables. At this stage, let us glance at two special problems that Part 1 is particularly concerned with, that is, [A] the total information in data and [B] joint graphical display. Let us now focus our attention on these problems.

3.6.1 What Does Total Information Mean?

We have extracted five components and the statistic δ indicates how much each component accounts for the total information. As far as δ is concerned, it looks as though the total value of 100 means that the information has exhaustively been extracted. This is a widely accepted view.

However, recall the point that each component requires two-dimensional space, one for rows and the other for columns. In this case, what does "$\delta_1 = 43\%$" mean? Since we need doubled multidimensional space, this statistic must also be re-evaluated in two-dimensional space associated with this component. How can we derive a statistic similar to δ for this expanded total space? What are the contributions of these five components in this doubled space?

As we will see later, the contingency table is a representation of packed information as we can infer from the space discrepancy angle θ. This statistic shows that each component from the contingency table requires two-dimensional space, and we need another statistic which shows the contribution of the contingency table component in doubled space.

Nishisato (1980) showed that the contingency table can be rewritten as a response-pattern table and that this latter table can reveal how each component from the contingency table can be represented in two-dimensional space. In this doubled 10-dimensional space of the Rorschach data, what is the proportion of information accounted for by each of the five components? We will discuss in Chap. 5 the derivation of this index, δT_{k1} and δT_{k2}, which show how much information two components associated with each contingency-table component k account for in the

Table 3.8 Order-0 to 4 approximations to the input table

		Fear	Anger	Depression	Love	Ambition	Security
Order-0	Bat	12.9	10.4	16.8	9.0	8.4	12.5
	Blood	3.3	2.7	4.3	2.3	2.2	3.2
	Butterfly	9.6	7.8	12.5	6.7	6.2	9.3
	Cave	5.0	4.0	6.5	3.5	3.2	4.8
	Clouds	9.6	7.8	12.5	6.7	6.2	9.3
	Fire	3.5	2.8	4.6	2.4	2.3	3.4
	Fur	7.0	5.7	9.1	4.9	4.6	6.8
	Mask	3.3	2.7	4.3	2.3	2.2	3.2
	Mountains	5.2	4.2	6.7	3.6	3.4	5.0
	Rocks	2.0	1.6	2.6	1.4	1.3	2.0
	Smoke	1.7	1.3	2.2	1.2	1.1	1.6
Order-1	Bat	24.4	15.2	23.5	−0.6	4.7	2.7
	Blood	7.0	4.2	6.4	−0.7	1.0	0.1
	Butterfly	−4.6	1.9	4.2	18.5	10.8	21.3
	Cave	7.3	5.0	7.9	1.5	2.5	2.8
	Clouds	12.4	8.9	14.1	4.3	5.3	6.9
	Fire	5.4	3.6	5.6	0.9	1.7	1.8
	Fur	0.0	2.8	5.0	10.7	6.8	12.7
	Mask	3.5	2.8	4.4	2.1	2.1	3.0
	Mountains	2.9	3.3	5.4	5.4	4.1	6.9
	Rocks	1.7	1.5	2.5	1.7	1.4	2.2
	Smoke	2.9	1.8	2.9	0.2	0.7	0.6
Order-2	Bat	25.1	16.0	23.1	0.9	1.2	3.7
	Blood	7.4	4.6	6.2	0.1	−0.9	0.7
	Butterfly	−3.2	3.5	3.2	21.5	3.8	23.2
	Cave	6.8	4.4	8.2	0.5	4.9	2.1
	Clouds	12.7	9.2	14.0	4.8	4.1	7.2
	Fire	5.7	4.0	5.4	1.6	0.0	2.3
	Fur	0.2	3.0	4.9	11.1	5.9	12.9
	Mask	3.5	2.7	4.5	2.0	2.3	3.0
	Mountains	0.3	0.3	7.2	−0.2	17.2	3.3
	Rocks	1.6	1.4	2.5	1.5	1.9	2.1
	Smoke	2.9	1.9	2.8	0.2	0.6	0.6
Order-3	Bat	27.0	17.3	18.9	1.4	2.1	3.3
	Blood	9.2	6.0	1.9	0.6	0.0	0.2
	Butterfly	−1.7	4.6	−0.3	21.9	4.6	22.8
	Cave	4.8	2.9	12.9	−0.1	3.9	2.6
	Clouds	5.9	4.3	29.2	3.1	0.8	8.7
	Fire	7.8	5.5	0.7	2.2	1.0	1.8
	Fur	−0.4	2.6	6.2	10.9	5.6	13.0
	Mask	2.8	2.2	6.0	1.9	2.0	3.1

(continued)

Table 3.8 (continued)

		Fear	Anger	Depression	Love	Ambition	Security
	Mountains	1.8	1.4	3.8	0.2	17.9	2.9
	Rocks	1.9	1.6	2.0	1.5	2.0	2.0
	Smoke	3.8	2.5	0.8	0.5	1.0	0.4
Order-4	Bat	32.9	9.9	18.5	2.4	2.1	4.2
	Blood	10.0	5.0	1.9	0.7	0.0	0.4
	Butterfly	0.3	2.2	−0.4	22.2	4.6	23.1
	Cave	7.1	0.0	12.7	0.3	3.9	2.9
	Clouds	2.1	9.1	29.4	2.4	0.8	8.1
	Fire	5.0	9.0	0.9	1.7	1.0	1.4
	Fur	−0.5	2.7	6.2	10.9	5.6	13.0
	Mask	3.0	2.0	6.0	1.9	2.0	3.1
	Mountains	2.1	1.0	3.7	0.3	17.9	3.0
	Rocks	0.0	4.0	2.1	1.2	2.0	1.7
	Smoke	1.0	6.0	1.0	0.0	1.0	0.0

Table 3.9 Statistic δ, exhaustive statistic δT, and angle

Components	1	2	3	4	5
δ	43%	23%	16%	12%	7%
δT_{k1}	86%	75%	71%	68%	63%
δT_{k2}	14%	25%	29%	32%	37%
θ (degree)	47°	60°	66°	69°	74°

expanded space. The formulas were derived from the relations between the two data formats (Nishisato and Sheu 1980).

$$\text{Main Component:} \quad \delta T_{k1} = 100(\frac{\rho_k + 1}{2})\%$$

$$\text{Supplementary:} \quad \delta T_{k2} = (100 - \delta T_{k1})\%$$

For the current example of five components from the contingency table, we can list the values of this statistic for the five components together with relevant statistics as in Table 3.9.

The above table indicates that, for example, the component 1 from the contingency table represents 86 per cent of information on this dimension, meaning that we need the second dimension to explain the remaining 14 per cent of its total information. Similarly, the component 5 from the contingency table contains only 63 per cent of its total information, leaving 37 per cent of information unaccounted for, that is, we need the second supplementary dimension to explain the remainder of 37 per cent of information. Thus, this new statistic tells us that the quantification analysis of the contingency table leaves a lot of information unaccounted for because it deals with only half the multidimensional space.

3.6.2 What is Joint Graphical Display

It is now clear that the joint graphical display of rows and columns cannot be depicted in the space associated with the contingency table, but must be in doubled multidimensional space. In other words, the joint graphical display of the Rorschach data requires 10-dimensional space, rather than the currently popular practice of five-dimensional space. How to expand the space and determine the coordinates in the doubled space is the most important focus of Part 1.

3.7 Currently Popular Methods for Graphical Display

With this much preliminary information, let us now look at the present practice of joint graphical display. This overview will primarily shed light on unsatisfactory aspects of the current practice, namely symmetric scaling (French plot), non-symmetric scaling, and CGS scaling. Pay attention to the fact that all the three methods deal with half the necessary space, namely five-dimensional space for our Rorschach data, rather than 10-dimensional space.

3.7.1 French Plot or Symmetric Scaling

In this plot, the graph is drawn, using principal coordinates $\rho_1 y_{i1}$ of rows and principal coordinates $\rho_1 x_{j1}$ of columns in the same space. However, since there is space discrepancy between row space and column space (i.e., the row-column correlation is not 1), each component requires two-dimensional space, that is, a two-dimensional graph for each component. Yet the symmetric scaling puts them into a single graph, calling it a joint graph. This is correct only when the row-column correlation is 1, the case which we do not typically observe in practice. Therefore, our verdict is that symmetric scaling offers at best only an approximation to the correct joint graphical display. The correct joint graph of *each component* must be a two-dimensional graph.

Thus, it is important to realize that symmetric scaling overlays the two graphs by ignoring the space discrepancy, and therefore our conclusion is that symmetric scaling is only an approximation to the correct joint graph.

3.7.2 Non-symmetric Scaling (Asymmetric Scaling)

The non-symmetric graph of component 1 is the plot of principal coordinates $\rho_1 y_{i1}$ of rows and standard coordinates x_{j1} of columns, or the plot of standard coordinates y_{i1} of rows and principal coordinates $\rho_1 x_{j1}$ of columns. This method appears logical at a first glance because it employs the projection operator to reduce the two-dimensional configuration to the unidimensional configuration. This is what the supporters of non-symmetric scaling say.

However, this method is logically and numerically wrong because this projection is utterly meaningless. As briefly mentioned earlier, standard coordinates are not the quantities to depict data structure because each component is standardized to a constant norm for every dimension, thus stripping off the unique information each component carries.

Recall that non-symmetric scaling starts with two sets of standard coordinates and then projects one set onto the data space, leaving the other set as is. Thus, the idea of projection is correct, but it must be applied to the sets of principal coordinates.

3.7.2.1 Revised Non-symmetric Scaling

In the context of meaningful projection (i.e., within the context of data structure), the correct method must start with principal coordinates of both row variables and column variables. Thus, if we wish to reduce doubled multidimensional space to the original space, the concept of projection should be applied to the two sets of principal coordinates: in other words, plot $(\rho_k y_{ik}, \rho_k^2 x_{jk})$ or $(\rho_k^2 y_{ik}, \rho_k x_{jk})$.

This is logically correct, but the only problem is that the two sets of graphed coordinates have different norms. This is completely against the original aim of joint graphical display of rows and columns on equal footing. This revised non-symmetric scaling is logically correct, but lacks the comparability between rows and columns—one set has a smaller norm than the other.

3.7.3 Comparisons

Because these two methods of joint graphical display are well known, we will show a very simple example to demonstrate their differences. Let us use a small subset of the Rorschach data by choosing only two Rorschach inkblots (Bat, Butterfly) and six moods (Fear, Anger, Depression, Love, Ambition, and Security) (Table 3.10). This data set yields only one component, which will make the comparison of the two graphical methods easy.

Quantification yields one component with eigenvalue 0.68 and singular value 0.83. This means that the discrepancy between row space and column space is 33 degrees, which is the arc cosine of the singular value. Let us keep in mind that the

Table 3.10 Two rorschach inkblots under six moods

	Induced moods					
Inkblots	Fear	Anger	Depression	Love	Ambition	Security
Bat	33	10	18	1	2	6
Butterfly	0	2	1	26	5	18

Table 3.11 Standard and principal coordinates

Rorschach	Principal coordinates	Standard coordinates
Bat	−0.71	−0.86
Butterfly	0.96	1.16
Fear	−0.86	−1.04
Anger	−0.52	−0.63
Depression	−0.75	−0.92
Love	1.09	1.31
Ambition	0.58	0.71
Security	0.65	0.79

axes for rows and columns are separated. The principal and standard coordinates of the two inkblots and the six moods are as shown in Table 3.11.

Note that the only quantities that can describe the data structure are principal coordinates, and that principal coordinates are standardized or normed to the same constant for all the components and as such standard coordinates no longer contain information about data structure. For this data set, the joint graphical displays by the two traditional methods are as follows:

A:**symmetric scaling.** Plot principal coordinates of both rows and columns.
B:**non-symmetric scaling (1).** Plot principal coordinates of rows and standard coordinates of columns.
C:**non-symmetric scaling (2).** Plot standard coordinates of rows and principal coordinates of columns.

Thus, the coordinates we will use for the three graphs are as shown in the table of coordinates. When we use these coordinates and draw graphs, we should note that non-symmetric scaling uses standard coordinates, which do not carry any information about the actual locations of the corresponding data points. So, we can immediately discard non-symmetric scaling graphs.

How about the plot under symmetric scaling, the popular French plot? We use principal coordinates, which is correct, but we must reject the plot, too, because the row weights lie along one axis and the column weights along another in such a way that the two axes cross the origin with the angle of 33 degrees! The conclusion is that the correct graph of a single component must be two-dimensional.

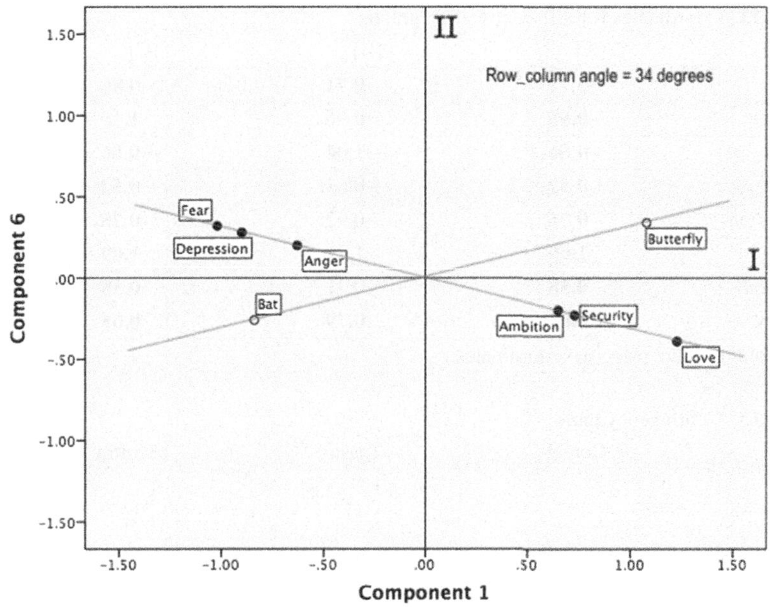

Fig. 3.5 Mathematically correct graph: rational 2-D graph

No matter what, it looks as though the inkblot Bat is close to the moods Fear, Anger, and Depressions, and the inkblot Butterfly is close to Love, Ambition, and Security. Although this graph looks reasonable and makes sense, we cannot ignore the fact that the row space and the column space are separated by an angle of 33 degrees! Thus, the symmetric scaling or French plot does not give us a correct graph either.

3.7.4 Rational 2-D Symmetric Plot

Using this discrepancy angle, we can draw a logically correct joint plot as we see in Fig. 3.5. Let us call this two-dimensional approach a rational graph with two axes, one for rows and the other for columns with the axes crossing at the origin with an angle of 33 degrees. Please note that this two-dimensional graph offers an exact geometric description of the relations of row variables and column variables. We will see the proof of this assertion in Chap. 5.

Table 3.12 Coordinates for different scaling graphs

Scaling	[A]	[B]	[C]
Bat	**−0.71**	**−0.71**	−0.86
Butterfly	**−0.96**	**−0.96**	1.16
Fear	**−0.86**	−1.04	**−0.86**
Anger	**−0.52**	−0.63	**−0.52**
Depression	**−0.75**	−0.92	**−0.75**
Love	**1.09**	1.31	**1.09**
Ambition	**0.58**	0.71	**0.58**
Security	**0.65**	0.79	**0.65**

Note: Boldfaced are principal coordinates

Table 3.13 Contingency table

	Cereal	Toast	Muffin
Coffee	2	1	1
Tea	1	2	1

3.7.5 CGS Scaling

Carroll et al. (1986) proposed an alternative method for arriving at a method for joint graphical display. Given a contingency table, the CGS scaling re-formats the contingency table into the form of an incidence table where the *rows* and the *columns* of the contingency table together will *occupy the rows of the incidence table*, and the new columns will consist of response combinations. To illustrate this transformation, let us use a small example in which the data are collected by asking subjects two multiple-choice questions:

- Q1: Do you prefer coffee to tea? (yes, no)
- Q2: Which do you like best for your breakfast? (cereal, toast, muffin)

Suppose that the data were obtained from 5 people as shown in Table 3.12. In the CGS scaling, this contingency table is represented as an incidence table (Table 3.13), where the crucial difference from the contingency table is that the rows and the columns of the contingency table now constitute the rows of the incidence table.

On the basis of the Young-Householder theorem (1938), therefore, the quantification of the incidence table yields quantities for both drinks and breakfast choices in common space. The problem of the traditional symmetric scaling was that the drinks in the rows and the breakfast choices in the columns of the contingency table are not in the same space. Thus, rewriting the contingency table to the form of the incidence table, the perennial problem of joint graphical display of space discrepancy can be solved. This was the gist of the CGS scaling, and this approach could have solved the perennial problem of joint graphical display. But, it did not!

From the previous discussion, we know that the contingency table of this example yields only one component, while the incidence table yields three components. An interesting observation is that the standard coordinates of the contingency table are exactly the same as the corresponding elements of the first component of the incidence table. In other words, so long as we compare only the first components from the contingency table of this example and its incidence table, there is no difference between the two formats—their principal coordinates are different because their first eigenvalues are different.

Then, what is the new contribution of the CGS scaling? The answer is that there is nothing new. The proponents of the CGS scaling have never discussed what the other two components from the incidence table are! This should have been the most important part of their proposal, but they never looked at these extra components. The opponent (Greenacre 1989) was neither aware of the importance of these extra components!

It was ironic that 6 years before the CGS scaling was proposed, Nishisato (1980) discussed those extra components from the incidence table. Later we will find out the importance of these extra components from the incidence format in solving the perennial problem of joint graphical display. In retrospect, it was too bad that neither the proponents nor the opponent ever considered the nature of those extra components.

The elements of the incidence table are either 1 or 0, and as such Greenacre (1989) considered that the quality of derived metrics, associated with them, was weak. He may be right on this point, but he cannot dismiss the CGS scaling because of that. Since Greenacre's criticisms (1989) of the CGS scaling were directed on the only components from the contingency table, it is no use to look at his criticisms of the CGS scaling, or the proponents' justifications (Carroll et al. 1987, 1989). Their arguments were not on the right track, and thus had no merits.

Keep in mind that the key to solving the perennial problem of joint graphical display resides in the extra components obtained from the incidence table. When this crucial point was explained in Nishisato's draft papers in 1989 and 2018, the editors of the two journals did not see the importance of his view and rejected his papers for publication in their journals. How frustrating has it been that the editors of the two well-known journals did not see the central issue of the problem! Three researchers encouraged him to advance his theory of extra dimensions for joint graphical display, Yasumasa Baba in Japan, Eric Beh in Australia, and José Clavel in Spain.

The readers have already seen some discussion on doubled dimensions for correct Euclidean graphs in this chapter. We will see further vindication of Nishisato's theory in Chap. 5. We will then see that the incidence table format or the response-pattern format yields generally more than twice the components from the contingency table, and that the most difficult task is to identify which components will supplement the doubled multidimensional space for the correct joint graphical display. In this process of reduction, Nishisato (2019a) developed a new scheme of classification of multidimensional quantification space, in which total space is further decomposed into contingency space, dual space, pairwise sets of dual subspace, and residual space. His theory on space partitions will be fully discussed in Chap. 4.

As the final background for joint graphical display, let us discuss in the next section that the contingency table is an information-packed form of data representation, and that we must unfold the information in the table into a larger format to see how much more extra information one can extract from the data. In Chap. 5, we will see that the unpacking of the contingency table can easily be carried out by quantifying the response-pattern (incidence) table associated with the contingency table.

3.8 Joint Graphs and Contingency Tables

There seem to be two definite and logical approaches to the problem of joint graphical display. One is based on quantification of the contingency table by incorporating the angle of discrepancy between row space and column space, and the other is a logical extension of the CGS scaling through analysis of the response-pattern table. We will look at the first approach below, and the second approach later in Chaps. 4 and 5.

Our object is to find the coordinates for rows and columns of the contingency table in common space. The key to solving the problem is embedded in the statistic, called *space discrepancy angle*, which is given by $cos^{-1}\rho_k$. As we briefly discussed earlier, all the data are centred with respect to each multidimensional axis.

With this much background, Nishisato (2016, 2019a; 2019b) handled this problem as follows: the separation angle is θ_k, but since the distribution of variables is assumed centred, we can introduce row axis and column axis, separated from the *horizontal axis* by angle $\frac{\theta_k}{2}$ in one direction and the other axis by the same angle in the opposite direction. His formulas for two-dimensional coordinates for component k are therefore given as follows: For row i and column j of component k, the two-dimensional coordinates are given, respectively, by

$$\text{Row i:} \quad \{\rho_k y_{ik}, \rho_k y_{ik} \sin\frac{\theta_k}{2}\}; \quad \text{Column j:} \quad \{\rho_k x_{jk}, -\rho_k x_{jk} \sin\frac{\theta_k}{2}\}$$

Because of symmetry and two dimensions, Tamaki Yano (personal communication, 2021) calls this graph *Symmetric Biplot*. This is a device to introduce two dimensions for each component associated with the contingency table.

Table 3.14 Incidence table

Subject*								
Coffee	1	1	1	1	0	0	0	0
Tea	0	0	0	0	1	1	1	1
Cereal	1	1	0	0	1	0	0	0
Toast	0	0	1	0	0	1	1	0
Muffin	0	0	0	1	0	0	0	1

Note: * The columns represent eight subjects

Table 3.15 Main and supplementary principal coordinates

Component	1	1*	2	2*	3	3*	4	4*	5	5*
Bat	−0.70	−0.28	−0.16	−0.08	0.15	0.08	−0.34	−0.19	−0.08	−0.05
Blood	−0.87	−0.35	−0.35	−0.18	0.60	0.33	−0.18	−0.10	0.06	0.04
Butterfly	1.17	0.47	−0.44	−0.22	0.17	0.09	−0.15	−0.08	0.32	0.19
Cave	−0.37	−0.15	0.30	0.15	−0.44	−0.24	−0.34	−0.19	0.11	0.07
Clouds	−0.23	−0.09	−0.08	−0.04	−0.75	−0.41	0.30	0.17	0.13	0.08
Fire	−0.42	−0.17	−0.30	−0.15	0.63	0.34	0.59	0.33	0.07	0.04
Fur	0.78	0.31	−0.08	−0.04	−0.08	−0.04	0.01	0.01	−0.67	−0.40
Mask	−0.05	−0.02	0.04	0.02	−0.21	−0.11	−0.04	−0.02	0.02	0.01
Mountains	0.34	0.14	1.54	0.77	0.31	0.17	−0.04	−0.02	0.11	0.07
Rocks	0.11	0.04	0.15	0.08	0.13	0.07	0.70	0.40	−0.08	−0.05
Smoke	−0.57	−0.23	−0.05	−0.03	0.57	0.31	1.25	0.71	0.00	0.00
Fear	−0.87	0.34	−0.17	0.09	0.39	−0.21	−0.48	0.27	−0.03	0.02
Anger	−0.44	0.18	−0.23	0.12	0.35	−0.19	0.75	−0.42	−0.02	0.01
Depression	−0.39	0.16	0.09	−0.05	−0.68	0.37	0.02	−0.01	0.10	−0.06
Love	1.03	−0.41	−0.51	0.26	0.15	−0.10	−0.12	0.07	0.48	−0.29
Ambition	0.42	−0.17	1.27	−0.04	0.29	−0.12	−0.10	0.00	0.05	−0.03
Security	0.76	−0.30	−0.23	0.12	−0.09	0.05	−0.08	0.05	−0.47	0.28

Note: Asterisks indicate supplementary components

Using the above formulas, we can double the multidimensional space from five-dimensional space to 10-dimensional space for the Rorschach data. Table 3.14 shows the coordinates of all the components, that is, the original five components and the five supplementary components. These are the 10-dimensional coordinates used for the exact Euclidean graph that places rows and columns in common space.

In contrast, symmetric scaling is based on the five-dimensional principal coordinates as shown in Table 3.7, offering an approximation to the 10-dimensional graph. Just note that symmetric scaling is based on the use of information embedded in only half the total space.

In Chap. 5, we will show that those coordinates of the 10 components in Table 3.14 indeed place rows and columns in common space correctly. As we noted, the discrepancy angle, associated with the contingency table, increases as we go down from the first to the second component, the last component showing the largest discrepancy. This aspect of space discrepancy will also be revisited in Chap. 5, where we will derive a more direct way to obtain principal coordinates for rows and columns in common space (Table 3.15). Fig. 3.5 is a 2-dimensional graph for the first component from the contingency table, and this graph is essentially the same as the graph obtained by plotting the components 1 and 1* of Table 3.15. Later we will find out that the ten-component principal coordinates are what we are looking for as the solution to the perennial problem of joint graphical display.

In Chap. 5, we will discuss much more about discrepancies between row space and column space, and we will introduce an index to show how well the traditional symmetric scaling represents row and column variables. We will then be able to assess how good French plot or symmetric scaling is.

3.8.1 A Theorem on Distance and Dimensionality

- When an n-dimensional graph is reduced to an (n-1)-dimensional graph, the distance between two points in the (n-1)-dimensional graph cannot become larger than the corresponding distance in the n-dimensional graph.

In other words, if the dimensionality of a graph is reduced for the purpose of joint display, the two points in the graph tend to get closer to each other. More boldly stated, the same graph in a smaller dimensional space would look better because data points tend to get closer to one another, often showing tighter clusters of variables. Look at our example of a one-component graph and the corresponding two-dimensional graph; it is clear that the distance between a row point and a column point in the one-dimensional graph is definitely closer than in the two-dimensional graph. In other words, the graph by symmetric scaling looks better than the graph by the rational approach, that is, the incorrect one looks better than the correct one.

This suggests a warning that whether the data configuration is interpretable or not alone cannot be used to determine how many dimensions the data set needs. Just remember that the smaller the dimensionality used in the graph, the closer the data points look, and this happens in particular to the distances between rows and columns.

The above observation provides a timely warning that the French plot tends to show a "better-looking" configuration of data points than the true configuration. This presents a cautious warning for the users of the French plot.

References

Bartlett, M. S. (1947). Multivariate analysis. *Journal of the Royal Statistical Society, Supplement, 9*, 176–190.

Bock, R. D. (1960). Methods and applications of optimal scaling. *The University of North Carolina Psychometric Laboratory Research Memorandum*, No. 25.

Carroll, J. D., Green, P. E., & Schaffer, C. M. (1986). Interpoint distance comparisons in correspondence analysis. *Journal of Marketing Research, 23*, 271–280.

Carroll, J. D., Green, P. E., & Schaffer, C. M. (1987). Comparing interpoint distances in correspondence analysis: A clarification. *Journal of Marketing Research, 24*, 445–450.

Carroll, J. D., Green, P. E., & Schaffer, C. M. (1989). Reply to Greenacre's commentary on the Carroll-Green-Schaffer scaling of two-way correspondence analysis solutions. *Journal of Marketing Research, 26*, 366–368.

Cronbach, L. J. (1951). Coefficient alpha and the internal structure of tests. *Psychometrika, 16*, 297–334.

Garmize, L. M., & Rychlak, J. F. (1964). Role-play validation of a socio-cultural theory of symbolism. *Journal of Consulting Psychology, 28*, 107–115.

Greenacre, M. J. (1989). The Carroll-Green-Schaffer scaling in correspondence analysis: A theoretical and empirical appraisal. *Journal of Marketing Research, 26*, 358–365.

Guttman, L. (1950). The basis for scalogram analysis. In S. A. Soufffer, et al. (Eds.), *Measurement and Prediction*. Princeton, N.J.: Princeton University Press.

Harman, H. H. (1960). *Modern Factor Analysis*. Chicago: University of Chicago Press.

Hirschfeld, H. O. (1935). A connection between correlation and contingency. *Cambridge Philosophical Society Proceedings, 31*, 520–524.

Horst, P. (1935). Measuring complex attitudes. *Journal of Social Psychology, 6*, 369–374.
Kalantari, B., Lari, I., Rizzi, A., & Simeone, B. (1993). Sharp bounds for the maximum of the chi-square index in a class of contingency tables with given marginals. *Computational Statistics and Data Analysis, 16*, 19–34.
Kendall, M. G. (1957). *A Course in Multivariate Analysis*. London: Charles Griffin and Company.
Lancaster, H.O. (1953) A reconsiliation of χ^2, considered from metrical and enumerative aspects. *Sakhya, 13*, 1–10.
Lord, F. M. (1958). Some relations between Guttman's principal components of scale analysis and other psychometric theory. *Psychometrika, 23*, 291–296.
Machine methods in scaling by reciprocal averages. *Proceedings, Research Forum*. Endicath, N.Y.: International Business Corporation. 35–39.
Nishisato, S. (1980). *Analysis of Categorical Data: Dual Scaling and Its Applications*. Toronto: The University of Toronto Press.
Nishisato, S. (1982). *Shitsuteki Data no Suryouka: Soutsuishakudo-ho to Sono Ohyou (Quantification of Qaulittative Data: Dual Scaling and its Applications)*. Tokyo: Asakura Shoten.
Nishisato, S. (1984). Dual scaling by reciprocal medians. *Proceedings of the 32nd Scientific Conference of the Italian Statistical Society*. Sorrento, Italy, 141–147.
Nishisato, S. (1987). Robust techniques for quantifying categorical data. In MacNeil, I.B. and Umphrey, G.J. (eds.), *Foundations of Statistical Inference*, 209–217. Dordrecht, The Netherlands: D. Reidel Publishing Company.
Nishisato, S. (1991). Standardizing multidimensional space for dual scaling. In the *Proceedings of the 20th Annual Meeting of the German Operations Research Society*. Hohenheim University, 584–591.
Nishisato, S. (1994). *Elements of Dual Scaling: An Introduction to Practical Data Analysis*. Hilsdale, N.J.: Lawrence Erlbaum Associates.
Nishisato, S. (2007). *Multidimensional Nonlinear Descriptive Analysis*. London: Chapman-Hall/CRC.
Nishisato, S. (2016). Quantification theory: Dual space and total space. Paper presented at the annual meeting of the Behaviormetric Society, Sapporo, Japan, p. 27 (In Japanese).
Nishisato, S. (2019a). Reminiscence: Quantification theory and graphs. *Theory and Applications of Data Analysis, 8*, 47–57. (in Japanese).
Nishisato, S. (2019b). Expansion of contingency space: Theory of doubled multidimensional space and graphs. An invited talk at the Annual Meeting of the Japanese Classification Society, Tokyo (in Japanese).
Nishisato, S. & Clavel, J.G. (2003). A note on between-set distances in dual scaling and correspondence analysis. *Behaviormetrika, 30*, 87–98.
Nishisato, S., & Nishisato, I. (1986). *Dual3 Users' Guide*. Toronto: MicroStats.
Nishisato, S., & Nishisato, I. (1984). *An Introduction to Dual Scaling*. Toronto: MicroStats.
Nishisato, S., & Nishisato, I. (1994). *Dual Scaling in a Nutshell*. Toronto: MicroStats.
Nishisato, S. & Sheu, W. (1980). Piecewise method of reciprocal averages for multiple- choice data of multiple-choice data. *Psyhometrika, 45*, 467–478.
Rao, C.R. (1952). *Advanced Statistical Methods for Biometric Research*. New York: Wiley.
Richardson, M. & Kuder, G. F. (1933). Making a rating scale that measures. *Personnel Journal, 12*, 36–40.
Thurstone, L. L. (1947). *Multiple Factor Analysis: A Development and Expansion of the Vector of Mind*. Chicago: The University of Chicago Press.
Torgerson, W.S. (1958). *Theory and Methods of Scaling*. New York: Wiley.
Williams, E. J. (1952). Use of scores for the analysis of association in contingency tables. *Biometrika, 39*, 274–289.
Yano, T. (2021). Personal comunication.
Young, G., & Householder, A. A. (1938). Discussion of a set of points in terms of their mutual distances. *Psychometrika, 3*, 19–22.

Chapter 4
Data Formats and Geometry

To understand why we need to double multidimensional space, we should look at some old studies. We used to carry out data analysis of the given data set, whatever the format of the data might be. There was typically no question raised about transforming the input data so as to carry out the specific task more effectively than otherwise. But, one should ponder if the given data format is informative enough to carry out desired analysis, or even appropriate for a specific task.

So, ask again if the given data format is appropriate for our analysis. This chapter is concerned with this question, and we will find out that additional attention to the data format is quite rewarding. More specifically speaking, we will find out that (1) the contingency table is a representation of tightly packed information, and that (2) it can be transformed in such a way that information hidden in the contingency table can be made more readily accessible. This transformation has the key to solve our perennial problem of the joint graphical display.

Six years before Carroll et al. (1986) proposed the so-called CGS scaling, Nishisato (1980) presented a way to unfold the contingency table into a form from which one could easily find the exact Euclidean coordinates for the rows and the columns of the contingency table in common space. Since his book was one of the mathematical exposition series of the University of Toronto Press, the relevant information he thoroughly presented may not have appealed to the eyes of investigators in quantification theory. But, the key information to solve the perennial problem of the joint graphical display was already described in the book, and one more step was needed for the CGS scaling to become a solution to the problem of the joint graphical display. This chapter will revisit and clarify his presentation with numerical examples so that the information relevant to solving the graphical problem may easily be understood. In this process, Nishisato (2019a) will explain his new theory of space partitions in Chap. 5.

S. Nishisato et al., *Modern Quantification Theory*,
Behaviormetrics: Quantitative Approaches to Human Behavior 8,
https://doi.org/10.1007/978-981-16-2470-4_4

4.1 Contingency Table in Different Formats

Nishisato (1980) presented three different ways of representing the contingency table and compared their geometric properties. Switching one format to another holds the key to expand our scope from the traditional space, called *contingency space*, to the space, called *dual space*, where quantified rows and columns can be accommodated in common space.

The key lies in representing the same data in three different formats (Nishisato 1980, Chap. 4). Consider his example, in which 13 subjects were asked the following two multiple-choice questions:

1. Do you smoke? (yes, no)
2. Do you prefer coffee to tea? (yes, not always, no)

First, we represent the data in the familiar contingency table (Table 4.1). Nishisato (1980) bridged simple correspondence analysis to multiple correspondence analysis, by treating the contingency table as the two-item multiple-choice data, and showed how the same data could be represented in three formats, namely contingency table, (1, 0) response-pattern table (incidence table), and the condensed response-pattern

Table 4.1 Contingency table format $\mathbf{F}$

	Drink		
Smoking?	Coffee	Not Always	Tea
Yes-smoking	3	2	1
Non-smoking	1	2	4

Table 4.2 Response-pattern table $\mathbf{F_a}$

Yes	No	Coffee	C or T	Tea
1	0	1	0	0
1	0	1	0	0
1	0	1	0	0
1	0	0	1	0
1	0	0	1	0
1	0	0	0	1
0	1	1	0	0
0	1	0	1	0
0	1	0	1	0
0	1	0	0	1
0	1	0	0	1
0	1	0	0	1
0	1	0	0	1

Table 4.3 Condensed response-pattern table $\mathbf{F}_{.}$

Yes	No	Coffee	C or T	Tea
3	0	3	0	0
2	0	0	2	0
1	0	0	0	1
0	1	1	0	0
0	2	0	2	0
0	4	0	0	4

table (table of distinct response patterns with frequencies). For the current example, the above data can be represented in these three formats (Tables 4.1, 4.2, and 4.3).

The contingency table $\mathbf{F}$ (Table 4.1) is 2×3, the (1,0) response-pattern table $\mathbf{F_p}$ (Table 4.2) is 13×5, and the condensed response-pattern table $\mathbf{F}_{.}$ (Table 4.3) is 6×5. If we are given one of the three formats, the other two formats can be derived from it. In this sense, the three formats look as though they are equivalent in terms of information content. However, not only the information content but also their mathematical structures are quite different.

4.2 Algebraic Differences of Distinct Formats

Among many interesting relations between different formats, Nishisato (1980) listed the following characteristics. Let us indicate the number of rows and that of columns by m and n, respectively, that is, $\mathbf{F}$ is $m \times n$. In this example, $\mathbf{F}$ is 2×3. Then, we see the following facts:

- The number of components from $\mathbf{F}$ is equal to *min(m, n) – 1*, that is, the (smaller of m and n) minus 1. Let us indicate this number as $N(\mathbf{F})$. In our example, $N(\mathbf{F}) = min(m, n) - 1 = min(2, 3) - 1 = 2 - 1 = 1$.
- The number of rows of $\mathbf{F_p}$ is the sum of the non-zero elements of $\mathbf{F}$, and the number of columns is $m + n$. In our example, $\mathbf{F_p}$ is 13×5. Notice that the rows and the columns of the contingency table are now arranged in the columns of $\mathbf{F_p}$. This is very important when we subject the data to quantification, for the Young-Householder theorem (Young and Householder 1938) guarantees that the quantification of $\mathbf{F_p}$ yields both rows and columns of the contingency table in the common Euclidean space.
- The number of components from $\mathbf{F_p}$, $N(\mathbf{F_p})$, is $m + n - 2$, that is, $N(\mathbf{F_p}) = m + n - 2$. In our example, $N(\mathbf{F_p}) = 3$. This format is essentially the same as the incidence matrix used by the proponents of the CGS scaling ("essentially" because the CGS scaling deals with the transpose of $\mathbf{F_p}$).
- The size of $\mathbf{F}_{.}$ is "the number of non-zero elements of $\mathbf{F_p}$" $\times$"$m + n$". In the current example, $\mathbf{F}_{.}$ is 6×5. Also in this format, the rows and the columns of $\mathbf{F_p}$

are represented in the columns of $\mathbf{F}_{.}$. Notice that $\mathbf{F}_{.}$ consists of the distinct response patterns with frequencies as elements. In other words, $\mathbf{F}_{.}$ is obtained from $\mathbf{F_p}$ by gathering identical response patterns into individual rows. $\mathbf{F}_{.}$, too, will yield row variates and column variates that span the common space as $\mathbf{F_p}$ does.

- Notice that $N(\mathbf{F_p}) = N(\mathbf{F}_{.})$. Both tables yield the same number of components, namely 3 in our example. In quantification, these two formats yield the same results with respect to the eigenvalues and the singular values and distinct elements of optimal weights.
- The above equivalence is extremely important for the current discussion. Mathematically, this equivalence can be explained by the *principle of distributional equivalence* (Benzécri et al. 1973). In the forced classification of dual scaling (Nishisato 1984a), this relation is referred to as the *principle of equivalent partitioning* and is effectively used in projecting data onto a specified subspace. Recall that Carroll, Green, and Schaffer used the transpose of $\mathbf{F_p}$, but this matrix is typically too large to use in practice. Thus, we will use only $\mathbf{F}_{.}$ in this chapter.
- It is of vital importance to note the following relation:

$$N(\mathbf{F}_{.}) \geq 2 \times N(\mathbf{F}),$$

where the equality holds when $m = n$. In our example, $N(\mathbf{F}_{.}) = 3$ and $N(\mathbf{F}) = 1$. We will find out that this relation holds the key to solve the perennial problem of the joint graphical display.

Notice that when rows and columns of the contingency table are placed in the same columns as in $\mathbf{F}_{.}$, we must at least double the dimensions of the quantification space to accommodate the rows and the columns in common space.

This is the crucial information to be used in arriving at the Euclidean coordinates for placing rows and columns in common space. Interestingly enough, neither the Carroll, Green, and Schaffer team nor Greenacre paid attention to this crucially important point, that is, *the quantification space must be expanded* to accommodate both row and column variates (Nishisato 1980). In other words, so long as the CGS proponents and Greenacre advanced their views *only* in the space associated with the contingency format **F**, their arguments were doomed to miss the most important point, that is, the centre of the problem. It is essential to realize that within the confine of the space associated with **F**, the traditional analysis and the CGS analysis provide *identical* results!

Carroll et al. (1986, 1987, 1989) and Greenacre (1989), therefore, totally missed this crucial point, and their arguments were out of the focus and meaningless. Interested readers are encouraged to read their papers and see how they insisted that their views were correct. To be exact, they did not discuss the problem of the joint graphical display.

Analysis of $\mathbf{F}_{.}$ alone, however, does not solve the perennial problem of the joint graphical display. For the Young-Householder theorem (Young and Householder 1938) to be properly applied to our problem, we need one more step to consider.

Before going further, however, let us observe the above relations numerically using the same example. Since we are interested in the relation between (rows, columns of **F**) and (columns of $\mathbf{F}_{.}$), we will restrict our attention to these relevant aspects, namely (smoking, non-smoking), (coffee, coffee or tea, tea) in our example.

Let us summarize the results of numerical analysis:

- The contingency table yields one component (C), while the response-pattern table has three components (R1, R2, R3) (see Tables 4.4 and 4.5).
- Notice that the singular value of **F** (say ρ) and the first eigenvalue of $\mathbf{F}_{.}$ (say ρ_{r1}^2) are related by

$$\rho = 2\rho_{r1}^2 - 1.$$

- Notice the distribution of information over three components (δ) of $\mathbf{F}_{.}$: Component 1 (R1) accounts for 49% of total information, R2 33%, and R3 18%.
- Cumulative percentages are from R1 to R3 49%, 82%, and 100%, respectively.
- The discrepancy between row space and column space of F is 63 degrees, that is, although **F** yielded one component, we need two-dimensional space to accommodate the "true" distribution of row variables and column variables (wait until we discuss the analytical results of $\mathbf{F}_{.}$).
- Please remember that the eigenvalue of the second component of $\mathbf{F}_{.}$ is 0.5. We will see that the corresponding standard and principal coordinates have a special pattern with an interesting implication. At this point, we simply mention that the

Table 4.4 Comparisons of two formats of data

Statistics	**F**	$\mathbf{F}_{.}$		
Eigenvalue ρ^2	0.21	0.73	0.50	0.27
Singular value ρ	0.46	0.85	0.71	0.52
δ	100	49	33	18
$\Sigma\delta$	100	49	82	100
Angle θ (degree)	63°			

Table 4.5 Standard and principal coordinates of two formats

Question	C-S	C-P	R1-S	R2-S	R3-S	R1-P	R2-P	R3-P
Smoking	**1.08**	0.50	**1.08**	0.00	−1.08	0.92	0.00	−0.56
Non-smoking	**−0.93**	−0.43	**−0.93**	0.00	0.93	−0.79	0.00	0.48
Coffee	**1.26**	0.58	**1.26**	−1.15	1.26	1.08	−0.81	0.66
Coffee or Tea	**0.17**	0.08	**0.17**	2.11	0.17	0.14	1.49	0.09
Tea	**−1.14**	−0.52	**−1.14**	−0.77	−1.14	−0.98	−0.54	0.59

Note: C = Contingency Table, R = Response-Pattern Table, S = Standard Coordinate
P = Principal Coordinate, Number 1, 2, 3 = Component 1, 2, 3

eigenvalue of 0.5 from $\mathbf{F}_{.}$ corresponds to the eigenvalue of 0 from **F**. What does this mean?

Let us now look at standard coordinates and principal coordinates associated with **F** and $\mathbf{F}_{.}$.

- Notice that the standard coordinates of the contingency table (C-S) in **boldface** and the standard coordinates of the first component of the response-pattern table (R1-S) in **boldface** (see Table 4.5) are identical! This is important to remember: as far as the common components from the two formats are concerned, they lead to the identical results!

4.3 CGS Scaling: Incomplete Theory

The historical arguments between Carroll et al. (1986) and Greenacre (1989) were over the common component(s) from **F** and $\mathbf{F}_{.}$, which as we have just seen are identical. How can one criticize one way or the other if the focus is on the common component? Their defence and Greenacre's attack are totally out of focus and the verdict is that there was no winner or loser.

4.4 More Information on Structure of Data

Let us continue our discussion on the relation of data formats and information.

- This equality of standard coordinates of the common components of **F** and $\mathbf{F}_{.}$ does not hold for the corresponding principal coordinates (C-P and R1-P) because the two eigenvalues are different, but they are *proportional*.
- The standard coordinates and the principal coordinates, associated with the eigenvalue of 0.5 (the second component of $\mathbf{F}_{.}$ in our example), show 0.00 for the two categories of smoking. In other words, this component has nothing to do with smoking. Thus, although the response-pattern table yielded three components, one of them does not contribute to the relation between the rows and the columns of the contingency table. Nishisato (1980) showed that the eigenvalue of 0.5 from $\mathbf{F}_{.}$ corresponds to the eigenvalue of 0 from the F, that is the case of no association between rows and columns.
- The above point indicates that the total number of components from $\mathbf{F}_{.}$ which capture the row-column relation of the contingency table is 2, that is, twice the number of component of **F**. This is the background for the theory of doubled multidimensional space. In the current example, the principal coordinates of rows and columns in common space are as shown in Table 4.6, and this is the answer to the perennial problem of joint graphical display, that is, these are the coordinates we have been looking for.

Table 4.6 Principal coordinates in common space

Question		Component 1	Component 3
1	Smoking	0.92	−0.56
	Non-smoking	−0.79	0.48
2	Coffee	1.08	0.66
	Coffee or Tea	0.14	0.09
	Tea	−0.98	0.59

Nishisato (1980) identifies more generally the following as differences between the two data formats.

- When we restrict our attention only to those components from the contingency table and the corresponding components from the response-pattern table (i.e., the first $(min(m, n) - 1)$ components of the response-pattern table (Note: in our example, the first component from each of the two formats as discussed above), there exist the following relations between the eigenvalues from the analysis of **F**, ρ_c^2, and those from the analysis of $\mathbf{F}_{.}$, ρ_f^2:

$$\rho_c^2 = (2\rho_f^2 - 1)^2, \text{ that is, } \rho_f^2 = \frac{\rho_c^4 + 1}{2}.$$

  Therefore, we obtain that

$$\rho_f^2 = 0.5 \text{ when } \rho_c^2 = 0.$$

  This means that when we analyse the response-pattern table and the eigenvalue of the component is 0.5, it is the case in which rows and columns are uncorrelated, that is, the component we would not be interested in so long as we seek the analysis of the row-column relations.
- For those corresponding components, there exist the following inequalities:

$$\rho_f^2 \geq \rho_c^2, \text{ equality holding when } \rho_f^2 = 1.$$

- Notice the relation that

$$1 \geq \rho_f^2 \geq 0.5 \text{ when } 1 \geq \rho_c^2 \geq 0.$$

- For those corresponding components, the standard coordinates associated with **F** are identical to the standard coordinates of $\mathbf{F}_{.}$. This is important, for as long as we look at all those corresponding components from the two formats, **F** and $\mathbf{F}_{.}$, the results are in essence identical, the only difference being in their singular values which make the two corresponding sets of principal coordinates different but proportional. As noted above, the arguments over the CGS scaling between the proponents and the opponent were over these common components, meaning that they had argued over something with no substance to argue about.

The above two relations are limited only to the first $(min(m, n) - 1)$ components.

In addition to those above, Nishisato (1993, see also 1994) lists the following points on analysis of $F_{.}$:

- The sum of the eigenvalues from the response-pattern table is equal to the average number of categories (response options) minus 1.

$$\Sigma\rho_{fk}^2 = \frac{\Sigma m_j}{n} - 1.$$

For our two-item case, this statistic is given by

$$\Sigma\rho_{fk}^2 = \frac{m+n}{2} - 1.$$

In our example, the two variables have 2 and 3 categories, so that the average number of response options (categories) is (3 + 2)/2 = 2.5. This minus 1 is 1.50, which is equal to the sum of the eigenvalues of the response-pattern table, namely, $0.73 + 0.50 + 0.27 = 1.50$.
- When data are from a **single** multiple-choice item with n response options, obtained from N respondents, where N is larger than $n - 1$, then $(n - 1)$ eigenvalues are all equal to 1.
- Nishisato (2016a) added the following to the list on analysis of $\mathbf{F}_{..}$. When the numbers of response options of two items are different (say m smaller than n), **each** of $(n - m)$ eigenvalues is equal to 0.5. This is related to the point (1) that these components correspond to the case in which the correlation between rows and columns is zero, and (2a) that the eigenvalues of a single multiple-choice item with categories are all equal to 1, and the point (2b) that in our case of two variables, the category coordinates of one variable are all 0, hence as a whole (2a and 2b), each eigenvalue in this case of two variables becomes 0.5, which is 1/2.
- The average of all the eigenvalues is also 0.5. In our example, $(3 - 2) = 1$, and there is one eigenvalue of 0.5. The average of the remaining eigenvalues is $(0.73 + 0.27)/2 = 0.5$. Thus, the average of all three eigenvalues is also 0.5.

Let us now use all these characteristics of quantification statistics to determine an exact set of Euclidean coordinates for row variables and column variables in common space, the task that the CGS scaling intended to do, but without success. In retrospect, it is too bad that Nishisato (1980) did not explore the application of these mathematical constraints to the problem of joint graphical display, but left the relevant information as an algebraic structure of $\mathbf{F}_{..}$. This structure immediately suggested that we must double multidimensional space to accommodate both row and column weights in common space.

Let us now apply the above discussion to solve the perennial problem of the joint graphical display. Now that we have all relevant information, the solution for the joint graphical problem is straightforward.

References

Benzécri, J. P. et al. (1973). *L'analyse des données: II. L'analyse des correspondances*. Paris: Dunod.

Carroll, J. D., Green, P. E., & Schaffer, C. M. (1986). Interpoint distance comparisons in correspondence analysis. *Journal of Marketing Research, 23*, 271–280.

Carroll, J. D., Green, P. E., & Schaffer, C. M. (1987). Comparing interpoint distances in correspondence analysis: A clarification. *Journal of Marketing Research, 24*, 445–450.

Carroll, J. D., Green, P. E., & Schaffer, C. M. (1989). Reply to Greenacre's commentary on the Carroll-Green-Schaffer scaling of two-way correspondence analysis solutions. *Journal of Marketing Research, 26*, 366–368.

Greenacre, M. J. (1989). The Carroll-Green-Schaffer scaling in correspondence analysis: A theoretical and empirical appraisal. *Journal of Marketing Research, 26*, 358–365.

Nishisato, S. (1980). *Analysis of Categorical Data: Dual Scaling and Its Applications*. Toronto: The University of Toronto Press.

Nishisato, S. (1984a). Forced classification: A simple application of a quantification technique. *Psychometrika, 49*, 25–36.

Nishisato, S. (1993). On quantifying different types of categorical data. *Psychometrika, 58*, 617–629.

Nishisato, S. (1994). *Elements of Dual Scaling: An Introduction to Practical Data Analysis*. Hilsdale, N.J.: Lawrence Erlbaum Associates.

Nishisato, S. (2019a). Reminiscence: Quantification theory and graphs. *Theory and Applications of Data Analysis, 8*, 47–57 (in Japanese).

Young, G., & Householder, A. A. (1938). Discussion of a set of points in terms of their mutual distances. *Psychometrika, 3*, 19–22.

Chapter 5
Coordinates for Joint Graphs

The aim of this chapter is to explain the theory of doubled multidimensional space for bi-modal analysis. In our attempt to derive coordinates for rows and columns of the contingency table, let us refresh our memories on the following points:

- The coordinates we seek are *principal coordinates*, not standard coordinates.
- Both rows and columns are equally treated in *bi-modal analysis*.
- *Expanded space* is needed to accommodate row and column variates in common space, which can be attained easily by quantifying the response-pattern format of the contingency table.

5.1 Coordinates for Rows and Columns

We have so far emphasized the bi-modal aspect of contingency-table analysis. Considering that rows and columns are generally not perfectly correlated, the space to accommodate both rows and columns of quantified data must be expanded, or more specifically, we must double our multidimensional space. In Chap. 3, we saw an example in which each of the 5 components required two-dimensional space, while the corresponding response-pattern table yielded more than 10 components. That example clearly indicates that the response-pattern format contains much more information than its contingency-table format, and that we must identify which 10 components constitute common space for both rows and columns of the contingency table. This has been discussed in Chap. 4 and we will extend this extra task in this chapter. We will then find the solution to our perennial problem of joint graphical display.

S. Nishisato et al., *Modern Quantification Theory*,
Behaviormetrics: Quantitative Approaches to Human Behavior 8,
https://doi.org/10.1007/978-981-16-2470-4_5

Although the example of the Rorschach data requires ten sets of coordinates for rows and columns, the proponents of the CGS scaling confined themselves only within the space of the contingency table, that is, five sets of coordinates. Greenacre's criticism of the CGS scaling was also confined within the same number of components as the one from the contingency table.

Using the same example of the Rorschach Responses and Moods, let us illustrate various steps to identify common Euclidean space for both rows and columns. We will start with the simplest situation in which the contingency table can be explained by one component, and then will gradually move towards more realistic situations which will then allow us to generalize the notion of doubled dimensionality as common space for rows and columns of the contingency table.

5.2 One-Component Case

As we can recall, the original data set of the Rorschach data from Garmize and Rychlak (1964) was slightly modified by eliminating five inkblots which had very small frequencies, and we saw the reduced data set in Chap. 3. For the current exposition, we will further reduce the data size to illustrate key points. First, let us consider only two Rorschach inkblots, Bat and Butterfly, together with the six induced moods from the original data. The two response formats $\mathbf{F}$ and $\mathbf{F}_{\cdot}$ are as shown in Tables 5.1 and 5.2.

When the contingency table (Table 5.1) is subjected to quantification, it yields one component as shown in Table 5.3. The chi-square for row-column independence is 83.13, with 5 degrees of freedom, which is significant at the 0.05 level.

If we use symmetric scaling, we obtain a unidimensional graph (Fig. 5.1). We clearly see that Rorschach inkblot "Bat" is closely associated with such moods as Fear, Anger and Depression, while Rorschach inkblot "Butterfly" is located close to such moods as Ambition, Security and Love. The results appeal to our common sense as "make sense" or interpretable. This is what symmetric scaling provides us.

However, the fact remains that the axis for row variates and the axis for column variates are separated by the angle of 34 degrees, as shown in Table 5.3. In other words, the axis for rows and the axis for columns should be as shown in Fig. 5.2 with the 34-degree angle between the two axes. Since the quantification is carried out with centred quantities, Fig. 5.2 is based on one axis with 17 degrees in one direction

Table 5.1 Two Rorschach inkblots under six moods

Inkblots	Induced moods					
	Fear	Anger	Depression	Love	Ambition	Security
Bat	33	10	18	1	2	6
Butterfly	0	2	1	26	5	18

Table 5.2 Response-pattern format

Bat	But	Fea	Ang	Dep	Lov	Amb	Sec
33	0	33	0	0	0	0	0
10	0	0	10	0	0	0	0
18	0	0	0	18	0	0	0
1	0	0	0	0	1	0	0
2	0	0	0	0	0	2	0
6	0	0	0	0	0	0	6
0	2	0	2	0	0	0	0
0	1	0	0	1	0	0	0
0	26	0	0	0	26	0	0
0	5	0	0	0	0	5	0
0	18	0	0	0	0	0	18

Notes But = Butterfly, Fea = Fear, Ang = Anger
Dep = Depression, Lov = Love, Amb = Ambition, Sec = Security

Table 5.3 Statistics and principal coordinates

Eigenvalue ρ^2	Singular Value ρ	Discrepancy Angle θ	$\omega_{str,1}$
0.68	0.83	34°	62
Bat	−0.71	Fear	−0.86
Butterfly	0.96	Anger	−0.52
		Depression	−0.76
		Ambition	1.09
		Security	0.58
		Love	0.65

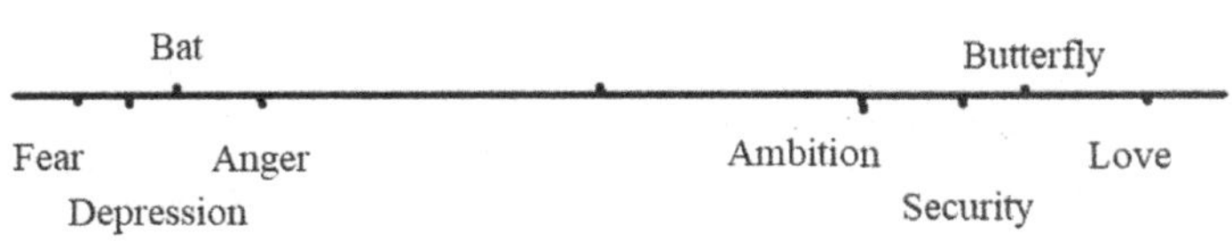

Fig. 5.1 Unidimensional scaling under symmetric scaling

and the other with 17 degrees in the opposite direction from the horizontal axis, as we discussed in Chap. 3. In a two-dimensional graph, the two row variables Bat and Butterfly should be allocated along one of the two axes and the six column variables Fear, Depression, Anger, Love, Ambition, and Security on the other axis. To find the exact coordinates of these variables, we would like to use the outcomes of the response-pattern analysis.

Quantification of the response-pattern table yields the results as summarized in Table 5.4.

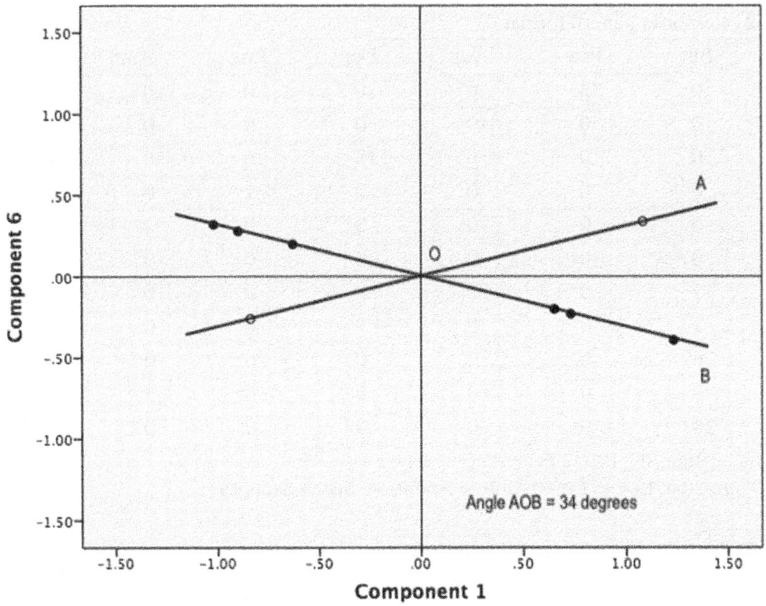

Fig. 5.2 Row axis and column axis

Table 5.4 Principal coordinates for the response-pattern Table

Component	1	2	3	4	5	6
Bat	−0.84	0.00	0.00	0.00	0.00	−0.26
Butterfly	1.08	0.00	0.00	0.00	0.00	0.34
Fear	−1.02	−1.33	0.43	0.08	−0.48	0.32
Anger	−0.63	1.33	1.80	−1.59	2.42	0.20
Depression	−0.90	1.40	−1.83	0.37	−0.23	0.28
Love	1.23	−0.30	−0.30	−0.60	0.54	−0.39
Ambition	0.65	0.79	1.08	3.29	0.43	−0.20
Security	0.73	−0.01	0.04	0.11	−1.15	−0.23
Eigenvalue ρ^2	0.91	0.50	0.50	0.50	0.50	0.09
Singular value ρ	0.95	0.71	0.71	0.71	0.71	0.30

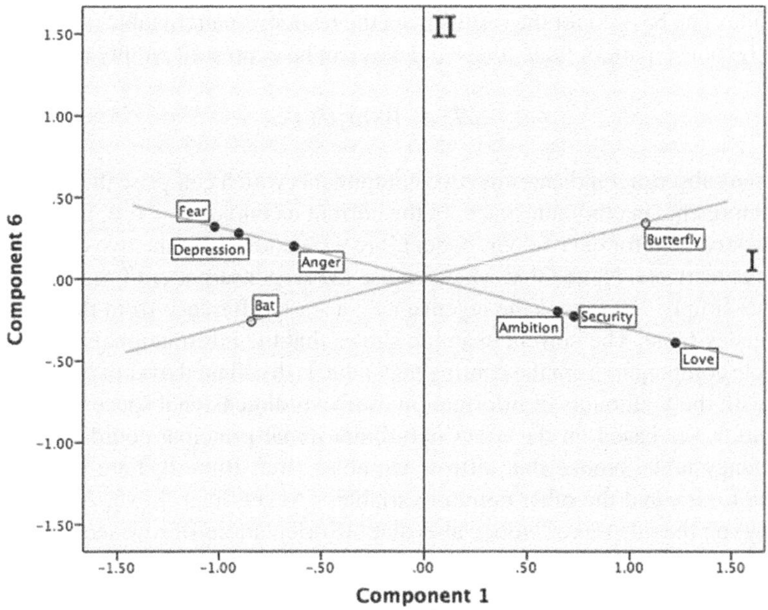

Fig. 5.3 Plot of components 1 and 6

To accommodate rows and columns of the contingency table (Table 5.1) in common space, we must identify two relevant components out of the six extracted components. This is the crucial aspect for the CGS scaling if it is to solve the joint graphical problem. As noted before, neither the proponents of the CGS scaling nor the opponent paid any attention to this issue.

As pointed out earlier, the chosen two-dimensional space must satisfy that the average eigenvalue is 0.5, and that each of the components associated with 0 correlation between rows and columns has the eigenvalue of 0.5. From the above table, those components with the eigenvalues 0.5 each are components 2, 3, 4, and 5. As mentioned earlier, these components 2, 3, 4, and 5 do not carry any information about the Rorschach inkblots (i.e., 0 weights on the inkblots), hence no information on the relation between Rorschach inkblots and moods. Once we exclude these components, there are two components, namely component 1 with the eigenvalue of 0.91 and component 6 with the eigenvalue of 0.09. These constitute our solution set. Notice that the average of the eigenvalues of these components in the solution set is $(0.91 + 0.09)/2 = 0.5$. Therefore, the two-dimensional coordinates which accommodate rows and columns in common space consist of those of components 1 and 6. This is our solution. If we plot those two components, we will get the same graph as in Fig. 5.3.

With respect to this pair of components, we have already proposed paired indexes (δT_{j1} and δT_{j2}) of exhaustiveness of their contributions to the doubled space. The same index can be used for the results from the response-pattern table. In the notation for the response-pattern table, these indexes can be expressed simply as

$$\delta T_u = 100\rho_u^2(\%)$$

where the subscript u indicates the two components which comprise the space for the two components in common space. In the current example, $u = 1, 6$, and the values of exhaustiveness for the two components are 91% and 9% for the two components 1 and 6, respectively. Notice that when we use the response-pattern format, this index becomes simply "100 times the eigenvalue", a stark difference from the case of the contingency table. The current example shows that the information associated with the single component from the contingency table is distributed over two-dimensional space with these amounts of information over two-dimensional space.

Figure 5.3 is based on the exact two-dimensional principal coordinates for our contingency table. Notice that all row variables (Bat, Butterfly) are located along one line (axis), and the other column variables (Anger, Fear, Love, Ambition, and Security) on the other axis. Notice also that the orientations of row and column axes of Fig. 5.2 are the same as those of Fig. 5.3.

Figure 5.3 is the solution to our perennial problem of joint graphical display! Thus, whether the graph by symmetric scaling is interpretable or not alone does not offer any justification that the unidimensional graph is correct, but for the current data, we need two-dimensional space, hence a two-dimensional graph. We should point out, however, that in the current example the unidimensional graph by symmetric scaling is comparatively good, considering that δT_1 is 91%.

It is important to remember again that the remaining components 2, 3, 4, and 5, each of which has the eigenvalue of 0.5, do not contribute to the correlation between rows and columns. Considering that the quantification of the contingency table is to capture the joint information of the rows and the columns, we define that these four components belong to *residual space*.

5.3 Theory of Space Partitions

The use of the response-pattern format and the contingency-table format has opened up our sight and grasp of quantification space. We are now ready to consider the structure of the quantification space we are dealing with in a much more meaningful way than before (Nishisato 2019, 2020).

We have so far used the term "dual space" without defining it, but it occupies the core of quantification theory, a reason why the name "dual scaling" was proposed. It is now opportune to take time away from numerical illustrations of analysis examples, and spend time to define different types of quantification space. In this way, we will be able to provide a solid framework for the quantification task. The following classification scheme of quantification space was proposed by Nishisato (2019, 2020):

- **Contingency Space:** This is the space used by symmetric scaling, where the dimensionality of space is given by the number of components associated with the contingency table, that is, $min(m, n) - 1$. In the above example, our contingency space is unidimensional. This is the space obtained by overlaying row space onto column space or vice versa, and this is what symmetric scaling deals with. In other words, even when row variates and column variates are not perfectly correlated, they are represented in contingency space as if they were perfectly correlated.
- **Dual Space:** This is the space which accommodates both row variables and column variables of the contingency table in common space. The dimension of dual space is twice that of contingency space. In the current example, dual space is two-dimensional. The average eigenvalue of the components in dual space is 0.5 (note: this applies only when two sets of categorical variables (i.e., row variates and column variates) are involved). The principal coordinates in dual space are the Euclidean coordinates for a perfect graph of row and column variates in common space, that is, the solution to the perennial problem of joint graphical display.
- **Pairwise Dual Subspace:** As noted before, the dimensionality of dual space is twice that of contingency space. When the dimensionality of dual space is more than 2, Nishisato (2019) discovered

  (1) that there are pairwise subspaces in dual space such that the average eigenvalue of each subspace is 0.5,

  (2) that each set of pairwise components yields a two-dimensional graph, and furthermore

  (3) that the pairwise components correspond to those two components, generated by incorporating the discrepancy angle between row space and column space of the contingency table (recall that we generated two components, one for rows and the other for columns when we extracted components from the contingency table, as we discussed in Chap. 3).
- Let us formally re-state that *each pair of the components, satisfying the average eigenvalue of 0.5, corresponds to one of the components in contingency space.* This opens up an interesting architecture of quantification theory: if you plot the two components in each dual subspace, it provides the true picture of what each component in contingency space should look like. This means that the exact principal coordinates for rows and columns of the contingency table can be obtained from the quantification of the contingency table, using space discrepancy angles as Nishisato suggested (recall Fig. 3.2 in Chap. 3). Note that the collection of these two-dimensional coordinates provides a solution to the perennial problem. In other words, the exact coordinates for both rows and columns in common space consist of the collection of the pairwise components from each of the contingency-table components. We now know, however, that a more direct way than that is to use principal coordinates from the analysis of the response-pattern table and choose those components which belong to dual space. In the current example, dual space consists of two components, and the two-dimensional graph of these two components is the exact Euclidean graph of the component obtained from the contingency table. As we will see in the next section, dual space has typically more than one pairwise subspaces, and the same argument will apply to each pairwise

dual subspace. When the contingency table yields 5 components, dual space is 10-dimensional space, which contains 5 pairwise subspaces, each depicting the true two-dimensional graph of a component associated with the contingency table.

- **Total Space:** This is the space which accommodates the total variation of the response-pattern table. When the number of rows is equal to the number of columns, total space is identical to dual space; when the number of rows is not equal to that of columns, total space has larger dimensionality than dual space. Again, there is a structural constraint that the average eigenvalue of total space is also 0.5. In the above example, the total space is six-dimensional, that is, the total number of components associated with the response-pattern table.
- **Residual Space:** When the number of rows is different from that of columns of the contingency table (e.g., the current example), the dimensionality of total space is larger than that of dual space, and the difference is called residual space. In residual space, either row variables or column variables have zero coordinates. It is important to note that the eigenvalue of *each* component in residual space is 0.5. In the previous chapter, we noted that this eigenvalue of 0.5 is obtained when the corresponding eigenvalue of the contingency table is 0, that is, when the correlation between rows and columns is 0. Another way of looking at this situation is to look at the eigenvalues of the response-pattern table for a single multiple-choice item, which are all equal to 1, and the residual space is exactly the same as analysing a single multiple-choice item because the categories of the other variable have no contributions, that is, zero coordinates. But, since there are two sets of variables (i.e., rows and columns) in the contingency table, each component in residual space has the eigenvalue equal to one half of 1, that is, 0.5.

In the current example, the table of the principal coordinates can be arranged in such a way that the first column is the coordinates of contingency space (C-space), and the first two columns are principal coordinates of all the variables in dual space (D-space). In this example, dual subspace consists of the same two components in dual space (see another example later where dual space consists of more than a pair of dual subspaces), the next four columns are principal coordinates of residual space (R-space: contributed only by moods in the current example), and all the six columns are coordinates of total space (T-space) as shown in Table 5.5. The *exact Euclidean coordinates* of the current example are those principal coordinates in dual space.

Recall that (1) dual space is the joint space of rows and columns of the contingency table, and that (2) total space is the joint space plus the residual space. Notice that in our example the residual space accommodates only the contributions of induced moods, and the contributions of the Rorschach responses to the residual space are all nil. In other words, all the relations between row variables and column variables are contained only in dual space. For this reason, we should conclude that the quantification theory of bi-modal analysis should be restricted only to dual space, an aspect that justifies the name "dual scaling".

5.4 Two-Component Case

Let us look at another example by adding one more variable of Rorschach inkblots, Mountains, to the first example. The data then consist of three Rorschach inkblots Bat, Butterfly, and Mountains, out of the 11 inkblots, and six induced moods Fear, Anger, Depression, Love, Ambition, and Security. The resultant 3×6 contingency table is given in Table 5.6, where the elements are joint frequencies. The corresponding condensed response-pattern table is as shown in Table 5.7.

Let us now look at the quantification results of the two formats. Table 5.8 shows the analysis of the contingency table: eigenvalues, singular values, and the angles between row space and column space for components. The chi-square for the row-column independence test yields the value of 151.41, with 10 degrees of freedom, which is significant at the level of 0.05.

Table 5.5 Analysis of the response-pattern table

Space	C-space					
	D-space	D-space				
			R-space	R-space	R-space	R-space
	T-space	T-space	T-space	T-space	T-space	T-space
Component	1	2	3	4	5	6
Bat	−0.84	−0.26	0.00	0.00	0.00	0.00
Butterfly	1.08	0.34	0.00	0.00	0.00	0.00
Fear	−1.02	0.32	−1.33	0.43	0.08	−0.48
Anger	−0.63	0.20	1.33	1.80	−1.59	2.42
Depression	−0.90	0.28	1.40	−1.83	0.37	−0.23
Love	1.23	−0.39	−0.30	−0.30	−0.60	0.54
Ambition	0.65	−0.20	0.79	1.08	3.29	0.43
Security	0.73	−0.23	−0.01	0.04	0.11	−1.15
Eigenvalue ρ^2	0.91	0.09	0.50	0.50	0.50	0.50
Singular value ρ	0.95	0.30	0.71	0.71	0.71	0.71

Table 5.6 Contingency table

Inkblots	Induced moods					
	Fear	Anger	Depression	Love	Ambition	Security
Bat	33	10	18	1	2	6
Butterfly	0	2	1	26	5	18
Mountains	2	1	4	1	18	2

Table 5.7 Response-pattern table

Inkblots			Moods					
Bat	But	Mtn	Fea	Ang	Dep	Lov	Amb	Sec
33	0	0	33	0	0	0	0	0
10	0	0	0	10	0	0	0	0
18	0	0	0	0	18	0	0	0
1	0	0	0	0	0	1	0	0
2	0	0	0	0	0	0	2	0
6	0	0	0	0	0	0	0	6
0	2	0	0	2	0	0	0	0
0	1	0	0	0	1	0	0	0
0	26	0	0	0	0	26	0	0
0	5	0	0	0	0	0	5	0
0	18	0	0	0	0	0	0	18
0	0	2	2	0	0	0	0	0
0	0	1	0	1	0	0	0	0
0	0	4	0	0	4	0	0	0
0	0	1	0	0	0	1	0	0
0	0	18	0	0	0	0	18	0
0	0	2	0	0	0	0	0	2

Notes But = Butterfly, Mtn = Mountain, Fea = Fear, Ang = Anger
Dep = Depression, Lov = Love, Amb = Ambition, Sec = Security

Table 5.8 Analysis of contingency table

	Component 1	Component 2
Eigenvalue ρ^2	0.62	0.38
Singular value ρ	0.79	0.62
Accounted for (δ)	62	38
$\Sigma\delta$	62	100
Discrepancy angle θ	42°	57°

Quantification of the contingency table yields two components, that is, contingency space is two-dimensional. This means that we need four-dimensional space to accommodate both rows and columns in common space (i.e., dual space).

The traditional two-dimensional symmetric graph is as shown in Fig. 5.2. One can interpret the configuration that Inkblot Bat is close to such moods as Depression, Anger, and Fear; Inkblot Butterfly is located near Love and Ambition; and Inkblot Mountains is close to Security. This two-dimensional graph is indeed easy to look at and immediately interpretable. How wonderful!

But, we must remember that the symmetric-scaling graph is an approximation to the four-dimensional configuration, and as we mentioned earlier the distances

Table 5.9 Analysis of the response-pattern table

Component	1	2	3	4	5	6	7
Eigenvalue ρ^2	0.895	0.81	0.50	0.50	0.50	0.19	0.105
Singular value ρ	0.95	0.90	0.71	0.71	0.71	0.44	0.33

between row variables and column variables in reduced space (in the current example, 2 dimensions) are depicted *closer* than they actually are! In other words, we can say that by presenting a four-dimensional graph in two dimensions, the relations between row variables and column variables look closer than they actually are! If one wants to be exact, therefore, the graph must be four-dimensional.

How can we depict a four-dimensional graph? We must develop a way to create a four-dimensional graph, which, however, is a problem for the future work. Thus, unless we develop an easy way to present a four-dimensional graph, the French plot of two dimensions in the current example can be an excellent practical compromise. We will look at possible alternatives in Chap. 6.

In the meantime, let us get back to our problem and consider it a little further. First, note that the response-pattern table yields seven components, and we must identify which of the four components are required for the exact Euclidean plot of the rows and the columns of the contingency table. The key statistics of the seven components are as shown in the table (Table 5.9).

The dimension of dual space is twice that of contingency space, that is, four. Again, the average eigenvalue of dual space must be 0.5. We should be reminded that the coordinates in dual space can also be obtained from the pairwise plots of individual components in contingency space, using the discrepancy angle. This means that dual space must include those components in the contingency table. Using this condition, the four components we are looking for are components 1, 2, 6, and 7 (see the average of 0.895, 0.811, 0.189, and 0.105 is 0.50). We can immediately identify a dual subspace consisting of component 1 and component 7 (the average of the eigenvalues of these two components is 0.5), and the other dual subspace consisting of components 2 and 6, of which the average eigenvalue is 0.5. Using the fact that each component in residual space has the eigenvalue of 0.5, we can easily identify those components in residual space, namely components 3, 4, and 5. The principal coordinates of the four dual-space dimensions are as summarized in Table 5.10. Remember that these are exact Euclidean coordinates for the current data.

Regarding dual subspaces, we should note that we use two components for symmetric scaling and that those two components are the ones associated with the two largest eigenvalues, that is, 0.895 and 0.81. As we can see, the second component in each subspace is comparatively minor (i.e., 0.105, 0.19) and they are the ones ignored in symmetric scaling. Here, we can understand why one wants to justify the use of symmetric scaling, and we finally see the rationale of symmetric scaling, revealing the ingenuity of French researchers. Table 5.10 indicates how much each of com-

Table 5.10 Exact Euclidean coordinates of the four-dimensional dual space

Component	1	2	6	7
Bat	−0.92	−0.39	−0.19	−0.31
Butterfly	1.19	−0.49	−0.23	0.41
Mountains	0.09	1.88	0.91	0.03
Fear	−1.10	−0.42	0.20	0.38
Anger	−0.66	−0.37	0.18	0.23
Depression	−0.83	0.00	0.00	0.28
Love	1.37	−0.63	0.31	−0.47
Ambition	0.29	1.97	−0.95	−0.10
Security	0.78	−0.45	0.22	−0.27
Eigenvalues	0.895	0.81	0.19	0.105

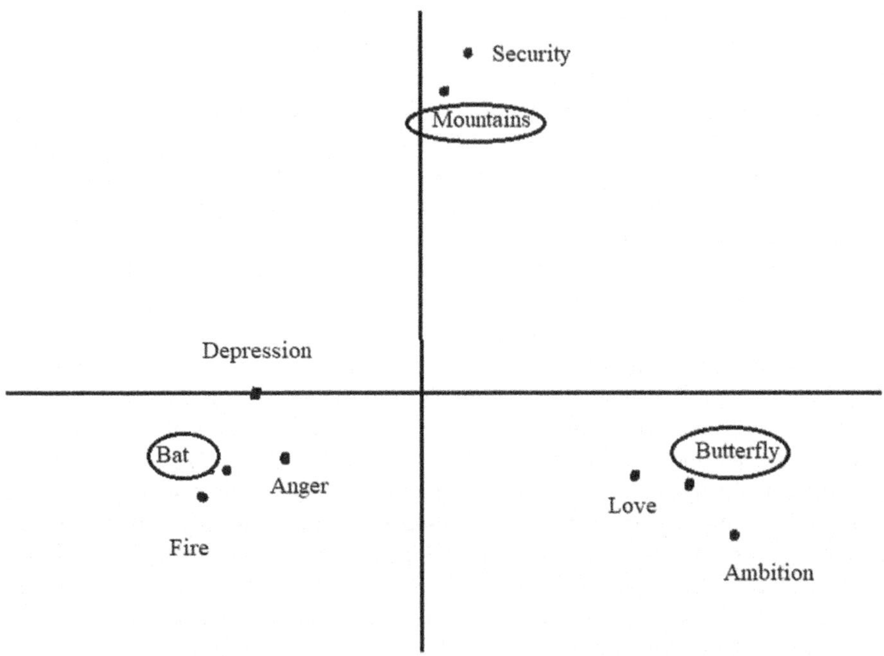

Fig. 5.4 Plot of components 1 and 2: symmetric scaling

ponent 1 and component 2 in Fig. 5.4 must be adjusted by an additional dimension. Although the eigenvalues of components 6 and 7 are comparatively very small, some of the elements of these components are still quite substantial, thus should not be ignored.

Keep in mind that the values of index δT_u are 89.5% and 81% for the two dual subspaces, or look at the corresponding space discrepancy angles, θ, of 42, and 57

Table 5.11 Principal coordinates of dual space: Subspace 1 and Subspace 2

Component	Subspace 1		Subspace 2	
	1	7	2	6
Bat	−0.92	−0.31	−0.39	−0.19
Butterfly	1.19	0.41	−0.49	−0.23
Mountains	0.09	0.03	1.88	0.91
Fear	−1.10	0.38	−0.42	0.20
Anger	−0.66	0.23	−0.37	0.18
Depression	−0.83	0.28	0.00	0.00
Love	1.37	−0.47	−0.63	0.31
Ambition	0.29	−0.10	1.97	−0.95
Security	0.78	−0.27	−0.45	0.22
Eigenvalues	0.895	0.105	0.81	0.19

degrees, respectively. In this way, we can understand how much information we are discarding by adopting symmetric scaling, namely 10.5% and 19%, or setting equivalently both angles of 42 and 57 degrees to 0!

As we noted earlier, if we plot two components in each dual subspace, row variables in the contingency table lie along one axis and the column variables on the other axis, and the angle between these two axes corresponds to the angle of discrepancy obtained in the analysis of the contingency table.

This demonstration clarifies the mathematical structure of quantification theory, namely dual space consists of pairwise subspace components, one in each pair with the eigenvalue greater than 0.5 and the other smaller than 0.5. All these components belong to dual space (Table 5.11).

The CGS scaling should have been proposed as a graphical method to plot all four components in dual space, but unfortunately this extension of the space was never discussed nor contemplated. If the proponents of the CGS scaling had realized the necessity of doubling the dimensionality of space, what graphical methods for doubled multidimensional space would they have proposed? It is too bad that those proponents of the CGS scaling missed the ample opportunity to demonstrate their well-known brilliance and expertise in proposing a graphical method for more than three-dimensional data.

Let us now look at the three-dimensional residual space, consisting of components 3, 4 and 5 (Table 5.12). We should note that the three Rorschach inkblots have all zero values, indicating that their entire contributions are accounted for by those four components in dual space, thus no more information is left in the residual space. Therefore, we can conclude that *although we extracted seven components, quantification theory should be concerned with only four components in dual space.*

Two merits in dealing with the response-pattern table are first that we can partition the total space into contingency space, dual space, and residual space, and secondly

Table 5.12 Principal coordinates of three-dimensional residual space

Component	Space		
	3	4	5
Bat	0.00	0.00	0.00
Butterfly	0.00	0.00	0.00
Mountains	0.00	0.00	0.00
Fear	−0.01	−1.25	0.29
Anger	2.74	1.23	0.78
Depression	−1.48	1.75	−0.42
Love	−0.09	0.11	1.23
Ambition	0.20	−0.29	0.16
Security	−0.14	−0.31	−1.88
Eigenvalues	0.50	0.50	0.50

that we can identify residual space which the traditional analysis of the contingency-table format fails to produce.

The last point is very important for research in general. There are many studies in which one of the two variables is dichotomous. In this case, no matter how many categories the other variable may have, dual space is two-dimensional and the contributions of many categories of the other variable to multidimensional space will be totally ignored, for we cannot further decompose the data in a meaningful way. The relations between the two categorical variables are mathematically restricted by the smaller number of categories of two variables.

From the strategic point of view, therefore, we should reduce residual space as much as possible. If $m = n$, the residual space is null. So, try to use the same numbers of categories for both rows and columns if possible.

Although we are not discussing the case of more than two categorical variables, the current point suggests how important it is to choose the same numbers of categories for all the variables so long as the number of choices of each variable is controllable.

It has been customary to interpret or to show in graph those components with relatively large eigenvalues. However, the current demonstration indicates that those components with the eigenvalues of 0.5 are in the residual space, that is, no importance in investigating the relations between row variables and column variables of the contingency table. This is a very important point that one should keep in mind when one wants to decide how many components to interpret. Any comparison of a component with another component which has the eigenvalue of 0.5 may be totally meaningless. Thus, never try to analyse all the components of eigenvalues greater than, for example, 0.3, which may include many components in residual space.

In leaving this section, let us summarize the decomposition of principal components in terms of different types of space (Table 5.13), where C-space = contingency space, D-space = dual space, DSub-1 is subspace 1 of dual space, DSub2 = subspace 2 of dual space, R-space = residual space, and T-space is total space. As noted

Table 5.13 Principal coordinates in different kinds of space

Space	C-space		C-space				
	D-space	D-space	D-Space	D-space			
	DSub-1	DSub-2	DSub-1	DSub-2			
					R-space	R-space	R-space
	T-space	T-space	T-space	T-space	T-space	T-space	T-space
Component	1	7	2	6	3	4	5
Bat	**−092**	−0.31	**−0.39**	−0.19	0.00	0.00	0.00
Butterfly	**1.19**	0.41	**−0.49**	−0.23	0.00	0.00	0.00
Mountain	**0.09**	0.03	**1.88**	0.91	0.00	0.00	0.00
Fear	**−1.10**	0.38	**−0.42**	0.20	−0.01	−1.25	0.29
Anger	**−0.66**	0.23	**−0.37**	0.18	2.74	1.23	0.78
Depression	**−0.83**	0.28	**0.00**	0.00	−1.48	1.75	−0.42
Love	**1.37**	−0.47	**−0.63**	−0.31	−0.09	0.11	1.23
Ambition	**0.29**	−0.10	**1.97**	−0.95	0.20	−0.29	0.16
Security	**0.78**	−0.27	**−0.45**	0.22	−0.14	−0.31	−1.88
Eigenvalue ρ^2	0.895	0.105	0.811	0.189	0.500	0.500	0.500

already, the exact Euclidean coordinates for the current data set are given by principal coordinates of dual space.

5.5 Three-Component Case

It is often useful to ask if the relations established in terms of one and two components would hold for a larger number of components. This question is typically answered when we extend the comparison to three components. Whenever we count on numerical verifications, this is the lesson one learns that if some relations hold for one, two, and three variables, one can typically generalize the relations discovered by then. To make sure that what we discovered so far are general enough, let us consider one more example by adding Clouds to the set of three Rorschach inkblots (Bat, Butterfly, and Mountains) of the previous example. Then the contingency table is 4×6 as shown in Table 5.14.

The main results of this contingency table are summarized in Table 5.15. The chi-square for statistical independence of rows and columns is 224.84, with 15 degrees of freedom. This value is significant at the 0.05 level.

Notice that the discrepancy angle of component 3 is 80 degrees, indicating that component 3 is better treated as two components, one for row variables and the other for column variables. This example shows a case in which three-dimensional symmetric scaling would be a very poor approximation to the true six-dimensional configuration.

The response-pattern table, corresponding to Table 5.14, is as shown in Table 5.16, and corresponding basic results are shown in Table 5.17. The principal coordinates of the eight components are as shown in Table 5.18.

On these results, we can summarize the following key points:

- From the contingency-table analysis, contingency space is three-dimensional. Therefore, dual space is six-dimensional. But, notice the substantial space discrepancy in component 3 ($\theta_3 = 80$ degrees). Thus, symmetric scaling of component 3 would be a very poor approximation to the configuration in dual space.
- From the analysis of the response-pattern table, the six-dimensional dual space consists of components 1, 2, 3, 6, 7, and 8.
- The above conclusion is reached with the facts (1) that the average eigenvalue of these six components is 0.5 and (2) that each of components 4 and 5 has the eigenvalue of 0.5, namely these two components belong to the two-dimensional residual space.
- Dual space can be further decomposed into subspaces 1, 2, and 3, consisting of pairs of components (1, 8), (2, 7), and (3, 6).
- Each subspace has the average eigenvalue of 0.5. This constraint means that the larger the eigenvalue of one of the two, the smaller the eigenvalue of the paired component, telling us that symmetric scaling approximation to the pairwise dimension deteriorates as we go down the components from the first contingency component to the last one.
- Residual space is two-dimensional. In this space, there are no contributions to these components by the inkblots, thus we can ignore the residual space for analysis.
- Each component in residual space has the eigenvalue of 0.5.

Table 5.14 Subset of the Rorschach data (Garmize and Rychlak 1964)

Inkblot	Induced moods					
	Fear	Anger	Depression	Love	Ambition	Security
Bat	33	10	18	1	2	6
Butterfly	0	2	1	26	5	18
Clouds	2	9	30	4	1	6
Mountains	2	1	4	1	18	2

Table 5.15 Analysis of contingency table

Component	1	2	3
Eigenvalue ρ^2	0.54	0.37	0.20
Singular value ρ	0.73	0.61	0.45
Accounted for (δ)	49	34	18
$\Sigma\delta$	49	82	100
Angle θ	57°	68°	80°

Table 5.16 Response-pattern table

Inkblots				Moods					
Bat	But	Clo	Mtn	Fea	Ang	Dep	Lov	Amb	Sec
33	0	0	0	33	0	0	0	0	0
10	0	0	0	0	10	0	0	0	0
18	0	0	0	0	0	18	0	0	0
1	0	0	0	0	0	0	1	0	0
2	0	0	0	0	0	0	0	2	0
6	0	0	0	0	0	0	0	0	6
0	2	0	0	0	2	0	0	0	0
0	1	0	0	0	0	1	0	0	0
0	26	0	0	0	0	0	26	0	0
0	5	0	0	0	0	0	0	5	0
0	18	0	0	0	0	0	0	0	18
0	0	2	0	2	0	0	0	0	0
0	0	9	0	0	9	0	0	0	0
0	0	30	0	0	0	30	0	0	0
0	0	4	0	0	0	0	4	0	0
0	0	1	0	0	0	0	0	1	0
0	0	6	0	0	0	0	0	0	6
0	0	0	2	2	0	0	0	0	0
0	0	0	1	0	1	0	0	0	0
0	0	0	4	0	0	4	0	0	0
0	0	0	1	0	0	0	1	0	0
0	0	0	18	0	0	0	0	18	0
0	0	0	2	0	0	0	0	0	2

Notes But = Butterfly, Clo = Cloud, Mtn = Mountain
Fea = Fear, Ang = Anger, Dep = Depression, Lov = Love
Amb = Ambition, and Sec = Security

Table 5.17 Analysis of the response-pattern table

Component	1	2	3	4	5	6	7	8
Eigenvalue ρ^2	0.87	0.81	0.72	0.50	0.50	0.28	0.19	0.13
Singular value ρ	0.93	0.90	0.85	0.71	0.71	0.53	0.44	0.36
Accounted for (δ)	22	20	18	13	13	7	5	3
$\Sigma\delta$	22	42	60	72	85	92	97	100

Table 5.18 Principal coordinates of eight components

Component	1	2	3	4	5	6	7	8
1	−0.91	−0.13	0.81	0.00	0.00	−0.50	−0.06	−0.35
2	1.38	−0.66	0.31	0.00	0.00	−0.19	−0.32	0.54
3	−0.43	−0.34	−1.35	0.00	0.00	0.83	−0.17	−0.17
4	0.52	2.18	−0.10	0.00	0.00	0.06	1.07	0.20
5	−1.10	−0.02	1.44	−0.41	−0.05	0.89	0.01	0.43
6	−0.60	−0.26	−0.36	2.72	1.57	−0.22	0.13	0.24
7	−0.67	−0.14	−1.10	−0.77	−0.26	−0.68	0.07	0.26
8	1.44	−0.84	0.24	−0.21	1.11	0.15	0.41	−0.56
9	0.73	2.22	0.00	0.07	0.11	0.00	−1.09	−0.29
10	0.76	−0.53	0.15	0.02	−1.80	0.10	0.26	−0.30

Table 5.19 Principal coordinates in different spaces

Space	{C	C	C}					
	{D	D	D	D	D	D}	{R	R}
	{D1	D1}	{D2	D2}	{D3	D3}	{R	R}
Component	1	8	2	7	3	6	4	5
1	−0.91	−0.35	−0.13	−0.06	0.81	−0.50	0.00	0.00
2	1.38	0.54	−0.66	−0.32	0.31	−0.19	0.00	0.00
3	−0.43	−0.17	−0.34	−0.17	−1.35	0.83	0.00	0.00
4	0.52	0.20	2.18	1.07	−0.10	0.06	0.00	0.00
5	−1.10	0.43	−0.02	0.01	1.44	0.89	−0.41	−0.05
6	−0.60	0.24	−0.26	0.13	−0.36	−0.22	2.72	1.57
7	−0.67	0.26	−0.14	0.07	−1.10	−0.68	−0.77	−0.26
8	1.44	−0.56	−0.84	0.41	0.24	0151	−0.21	1.11
9	0.73	−0.29	2.22	−1.09	0.00	0.00	0.07	0.11
10	0.76	−0.30	−0.53	0.26	0.15	0.10	0.02	1.80

The principal coordinates of all the components can be re-organized in terms of space structure as in Table 5.19, where C = contingency space, D = dual space, D1, D2, and D3 are dual subspaces 1, 2, and 3, respectively, and R = residual space. Total space is the sum of dual space and residual space, that is, eight-dimensional. The exact Euclidean space for rows and columns of the contingency table in dual space is six-dimensional as shown in Table 5.19.

5.6 Wisdom of French Plot

We have so far looked at most of the basic procedure for finding coordinates for rows and columns of the contingency table in dual space. In this process, we were aware that most researchers are interested in joint graphs, but a practical question is always there, that is, how to use this knowledge of exact coordinates when the number of components in dual space is greater than two. In this regard, remember that the dimensionality of dual space increases from 2 to 4, 6, 8, and so on in multiples of two, and unless we develop an easy, effective graphical method for more than two-dimensional space, how can we use our knowledge of multidimensional coordinates?

In contrast, the dimensionality for the contingency space, which the French plot deals with, increases from 1, 2, 3, 4, and so on, that is, the French plot deals with half the space. The question then is if its approximation to the correct joint graph is good enough. For this, we should examine the discrepancy angle θ_k, that is, $cos^{-1}\rho_k$ and the index δT_u.

In our previous example of the two-component case, we obtained four components and coordinates of nine variables (Table 5.11). In the current practice, it is usually the case that we plot combinations of two components, such as (1, 2), (1, 6), (1, 7), (2, 6), (2, 7), and (6, 7). Out of these, the most widely chosen combination is probably the combination of (1, 2), the graph of which was earlier presented in Fig. 5.4. An interesting point is that out of these six pairs, we use only pair (1, 2) in symmetric scaling: in contingency space, we consider only components 1 and 2 in this example. Thus, in this particular example, symmetric scaling would be appealing: we see three clusters in six-dimensional space, namely (1) inkblot Butterfly with moods Love and Ambition, (2) inkblot Bat with moods Anger, Depression, and Fire, and (3) inkblot Mountains with mood Security.

We have noted that symmetric scaling deals with only those components of which the eigenvalues are greater than 0.5 (the response-pattern table) or 0.0 (the contingency table). From the approximation point of view, this is an excellent choice, which can be called "French Wisdom". No wonder why symmetric scaling is still the number one choice of graphical methods.

However, in the above example, there are five more combinations of two components and the question is if there are any combinations of these which provide interesting information. This question is so far left unanswered.

Thus, for those who are interested in approximating the correct graph, the French plot sounds quite reasonable and appears very attractive. It is indeed a masterful compromise. But, please remember the thoughtful warning by Lebart et al. (1977) that the between-set distance (i.e., the distance between a row and a column) in the French plot is not exact. This is so because the French plot is based on half the space necessary to describe coordinates of data points correctly.

Our desiderata are to pursue a new strategy to develop higher dimensional graphs using correct coordinates in dual space, or else we must find an alternative to it, such as cluster analysis, a topic to be discussed in the next chapter.

Table 5.20 Analysis of the complete contingency table

Component	1	2	3	4	5
Eigenvalue ρ^2	0.46	0.25	0.17	0.13	0.07
Singular value ρ	0.68	0.50	0.41	0.36	0.27
Accounted for δ (%)	43	23	16	12	7
$\Sigma\delta$	43	66	82	93	100
Angle θ (degree)	47°	60°	66°	69°	74°
Symmetric $\omega_{sym,k}$	48	33	27	23	18

5.7 General Case

These demonstrations so far may be sufficient to explain how principal coordinates for rows and columns of the contingency table can be derived through the quantification of the response-pattern format. However, to look at the entire data set as shown in Table 3.1 may still be instructive. We should note that the entire data set in the contingency-table format is not large, and it is relatively easy to analyse it. As we have already seen, the entire contingency table is 11×6. In this example, we will see that the contingency table is indeed tightly information-packed as compared with the response-pattern format. Table 5.20 shows summary statistics of the analysis of the contingency table, and we will first follow the same procedure as we have seen before with smaller examples. The chi-square for row-column independence is 370.85, degrees of freedom = 50, and this is significant at the 0.05 level of confidence. There are in total five components, and their principal coordinates are as shown in Table 5.21. The investigator's first question is typically to decide how many components to consider for analysis, including the graphical display. For instance, since the first three components account for a respectable amount of 82% of the total information, we may decide to look at the first three components. As for the values of δT_u, we obtain, from the analysis of the response-pattern table, 86%, 75%, and 71% for the three pairs of components in dual space. The values of this index indicate that we miss 14, 25, and 29% of information by adopting three components from the contingency table: symmetric scaling would miss a lot of information by ignoring the supplementary components. Since the current demonstration is to discuss some theoretical aspects of joint graphical display, let us move on.

Table 5.21 Principal coordinates of five components

Component	1	2	3	4	5
Bat	−0.70	−0.16	0.15	−0.34	−0.08
Blood	−0.87	−0.35	0.60	−0.18	0.06
Butterfly	1.17	−0.44	0.17	−0.15	0.32
Cave	−0.37	0.30	−0.44	−0.34	0.11
Clouds	−0.23	−0.08	−0.75	0.30	0.13
Fire	−0.42	−0.30	0.63	0.59	0.07
Fur	0.78	−0.08	−0.08	0.01	−0.67
Mask	−0.05	0.04	−0.21	−0.04	0.02
Mountains	0.34	1.54	0.31	−0.04	0.11
Rocks	0.11	0.15	0.13	0.70	−0.08
Smoke	−0.57	−0.05	0.57	1.25	0.00
Fear	−0.87	−0.17	0.39	−0.48	−0.03
Anger	−0.44	−0.23	0.35	0.75	−0.02
Depression	−0.39	0.09	−0.68	0.02	0.10
Love	1.03	−0.51	0.15	−0.12	0.48
Ambition	0.42	1.27	0.29	0.00	0.05
Security	0.76	−0.23	−0.09	−0.08	−0.47

Let us direct our attention to the analysis of the response-pattern table. The biggest surprise, which we have not seen so far, is the size of the response-pattern table, namely (the number of non-zero elements in the contingency table)×(the number of rows + the number of columns). Our response-pattern table is 58×17! What a jump in the data size! See the response-pattern table as shown in Table 5.22.

From the researcher's point of view, the size of the response-pattern table may soon reach the practical limit as the data size increases. What can we do then? We have already seen one alternative: go back to the original contingency table and use the formulas discussed in Chap. 3. Namely, we calculate the second component for each contingency-table component as follows: for row i and column j of component k, the second coordinates are given, respectively, by

$$\text{For row i:} \qquad \left\{\rho_k y_{ik}, \rho_k y_{ik} \sin\frac{\theta_k}{2}\right\}$$

$$\text{For column j:} \qquad \left\{\rho_k x_{jk}, -\rho_k x_{jk} \sin\frac{\theta_k}{2}\right\}$$

.

The entire set of pairwise components can be used as coordinates of the rows and the columns of the contingency table in common space (dual space). In this way, we can handle the size problem of the response-pattern table rather easily and analysis is restricted within dual space. Go back to Chap. 3 for the relevant discussion.

Let us now look at the analysis of the response-pattern table. The total number of components is the total number of rows and columns of the contingency table minus 2, which is 15. The 15-dimensional space includes the residual space.

Quantification results from the response-pattern table is as summarized in Tables 5.23 and 5.24.

The results of this section suggest a number of important matters:

- So long as we are interested in the multidimensional decomposition of the association between the rows and the columns of the contingency table, we should ignore the information in the residual space, which corresponds to components 6, 7, 8, 9, and 10 (notice the corresponding eigenvalues are all equal to 0.5). Without these components, the principal components of the remaining components are as shown in Table 5.25. This tells us how important it is to equate the number of rows to that of columns as much as possible in data collection (i.e., the dimensionality of residual space is nil when the number of rows is equal to the number of columns).
- Those remaining components in Table 5.25 are arranged in the descending order of the eigenvalues. We should note that the first five components are what symmetric

Table 5.22 Response-pattern table of the entire data set

Ba*	Bl	Bu	Ca	Cl	Fi	Fu	Ma	Mt	Rc	Sm	Fe	An	Dp	Lo	Am	Se
33	0	0	0	0	0	0	0	0	0	0	33	0	0	0	0	0
10	0	0	0	0	0	0	0	0	0	0	0	10	0	0	0	0
1	0	0	0	0	0	0	0	0	0	0	0	0	18	0	0	0
1	0	0	0	0	0	0	0	0	0	0	0	0	0	1	0	0
2	0	0	0	0	0	0	0	0	0	0	0	0	0	0	2	0
6	0	0	0	0	0	0	0	0	0	0	0	0	0	0	0	6
0	10	0	0	0	0	0	0	0	0	0	10	0	0	0	0	0
0	5	0	0	0	0	0	0	0	0	0	0	5	0	0	0	0
0	2	0	0	0	0	0	0	0	0	0	0	0	2	0	0	0
0	1	0	0	0	0	0	0	0	0	0	0	0	0	1	0	0
0	0	2	0	0	0	0	0	0	0	0	0	2	0	0	0	0
0	0	1	0	0	0	0	0	0	0	0	0	0	1	0	0	0
0	0	26	0	0	0	0	0	0	0	0	0	0	0	26	0	0
0	0	5	0	0	0	0	0	0	0	0	0	0	0	0	5	0
0	0	18	0	0	0	0	0	0	0	0	0	0	0	0	0	18
0	0	0	7	0	0	0	0	0	0	0	7	0	0	0	0	0
0	0	0	13	0	0	0	0	0	0	0	0	0	13	0	0	0
0	0	0	1	0	0	0	0	0	0	0	0	0	0	1	0	0
0	0	0	4	0	0	0	0	0	0	0	0	0	0	0	4	0
0	0	0	2	0	0	0	0	0	0	0	0	0	0	0	0	2

(continued)

Table 5.22 (continued)

Ba*	Bl	Bu	Ca	Cl	Fi	Fu	Ma	Mt	Rc	Sm	Fe	An	Dp	Lo	Am	Se
0	0	0	0	2	0	0	0	0	0	0	2	0	0	0	0	0
0	0	0	0	9	0	0	0	0	0	0	0	9	0	0	0	0
0	0	0	0	30	0	0	0	0	0	0	0	0	30	0	0	0
0	0	0	0	4	0	0	0	0	0	0	0	0	0	4	0	0
0	0	0	0	1	0	0	0	0	0	0	0	0	0	0	1	0
0	0	0	0	6	0	0	0	0	0	0	0	0	0	0	0	6
0	0	0	0	0	5	0	0	0	0	0	5	0	0	0	0	0
0	0	0	0	0	9	0	0	0	0	0	0	9	0	0	0	0
0	0	0	0	0	1	0	0	0	0	0	0	0	1	0	0	0
0	0	0	0	0	2	0	0	0	0	0	0	0	0	2	0	0
0	0	0	0	0	1	0	0	0	0	0	0	0	0	0	1	0
0	0	0	0	0	1	0	0	0	0	0	0	0	0	0	0	1
0	0	0	0	0	0	3	0	0	0	0	0	3	0	0	0	0
0	0	0	0	0	0	4	0	0	0	0	0	0	4	0	0	0
0	0	0	0	0	0	5	0	0	0	0	0	0	0	5	0	0
0	0	0	0	0	0	5	0	0	0	0	0	0	0	0	5	0
0	0	0	0	0	0	21	0	0	0	0	0	0	0	0	0	21
0	0	0	0	0	0	0	3	0	0	0	3	0	0	0	0	0
0	0	0	0	0	0	0	2	0	0	0	0	2	0	0	0	0
0	0	0	0	0	0	0	6	0	0	0	0	0	6	0	0	0
0	0	0	0	0	0	0	2	0	0	0	0	0	0	2	0	0
0	0	0	0	0	0	0	2	0	0	0	0	0	0	0	2	0
0	0	0	0	0	0	0	3	0	0	0	0	0	0	0	0	3
0	0	0	0	0	0	0	0	2	0	0	2	0	0	0	0	0
0	0	0	0	0	0	0	0	1	0	0	0	1	0	0	0	0
0	0	0	0	0	0	0	0	4	0	0	0	0	4	0	0	0
0	0	0	0	0	0	0	0	1	0	0	0	0	0	1	0	0
0	0	0	0	0	0	0	0	18	0	0	0	0	0	0	18	0
0	0	0	0	0	0	0	0	2	0	0	0	0	0	0	0	2
0	0	0	0	0	0	0	0	0	4	0	0	4	0	0	0	0
0	0	0	0	0	0	0	0	0	2	0	0	0	2	0	0	0
0	0	0	0	0	0	0	0	0	1	0	0	0	0	1	0	0
0	0	0	0	0	0	0	0	0	2	0	0	0	0	0	2	0
0	0	0	0	0	0	0	0	0	2	0	0	0	0	0	0	2
0	0	0	0	0	0	0	0	0	0	1	1	0	0	0	0	0
0	0	0	0	0	0	0	0	0	0	6	0	6	0	0	0	0
0	0	0	0	0	0	0	0	0	0	1	0	0	1	0	0	0
0	0	0	0	0	0	0	0	0	0	1	0	0	0	0	1	0

**Note* Bl = blood, Bu = butterfly, Cl = clouds, Fi = fire, Ma = masks
Mt = mountains, Rc = rocks, Sm = smoke, Fe = fear, An = anger
Dp = depression, Lo = love,Am = ambition, Se = security

Table 5.23 Basic statistics

Component	ρ_k^2	ρ_k	δ	Cumδ	θ_k
1	0.84	0.92	11.20	11.20	23
2	0.75	0.87	10.00	21.21	30
3	0.71	0.84	9.42	30.63	33
4	0.68	0.82	9.05	39.68	35
5	0.63	0.80	8.46	48.14	37
6	0.5	0.71	6.67	54.80	45
7	0.5	0.71	6.67	61.47	45
8	0.5	0.71	6.67	68.14	45
9	0.5	0.71	6.67	74.80	45
10	0.5	0.71	6.67	81.47	45
11	0.37	0.60	4.88	86.35	53
12	0.32	0.57	4.28	90.63	55
13	0.29	0.54	3.92	94.54	57
14	0.25	0.50	3.33	97.67	60
15	0.16	0.40	2.13	100.00	66

scaling (French plot) deals with while ignoring the remaining five components, that is, those components which serve to correct the discrepancies between row space and column space. When we look at the sizes of the eigenvalues (those first five components are all greater than 0.5, while the remaining components for space adjustment were all smaller than 0.5), we can see the justification of the French plot as an approximation to the correct configuration by each of those dominant components.

- However, look at the values of the index δT_u, which clearly show that the results of the French plot leave a lot of information unaccounted for.
- The above observation clearly divides researchers into two categories, those who consider the practical aspect more important than accuracy, and those who must see the correct representation of data structure.

5.8 Further Considerations

We have solved the problem of multidimensional coordinates of both rows and columns of the contingency table in common space. But, how can we use this knowledge in practice? In other words, our problem of quantification is not over yet because our brains are typically unable to visualize more than three-dimensional configurations while our quantification results often require more than three-dimensional graphs. So, our next task is how to make good use of our newly acquired knowledge in practice. There are a few alternatives we can look at.

Table 5.24 Principal coordinates of total space

Co	1	2	3	4	5	6	7	8	9	10	11	12	13	14	15
Ba	−1.09	0.26	−0.34	−0.74	−0.24	−0.77	0	0	0	1.03	−0.18	−0.51	−0.22	0.15	0.44
Bl	−1.12	0.63	−1.17	−0.32	−0.17	0.25	0.74	−0.28	1.39	−3.31	−0.13	−0.22	−0.75	0.36	0.46
Bu	1.51	0.74	−0.45	−0.51	0.92	−0.07	0.14	0.20	0.11	0.17	0.70	−0.35	−0.29	0.43	−0.62
Ca	−0.47	−0.50	0.89	−0.83	0.13	2.60	−0.45	1.31	0.29	0.14	0.10	−0.57	0.57	−0.29	0.19
Cl	−0.26	0.17	1.53	0.57	0.49	−0.95	−0.12	0.10	−0.02	−0.55	0.37	0.39	0.98	0.10	0.11
Fi	−0.53	0.54	−1.20	1.39	0.23	0.61	−2.51	0.11	−2.06	−0.35	0.18	0.96	−0.77	0.31	0.22
Fu	1.12	0.12	0.20	0.19	−1.97	−0.12	0.00	0.33	−0.12	−0.16	−1.50	0.13	0.13	0.07	−0.46
Ma	−0.04	−0.06	0.43	−0.12	0.00	1.38	1.61	−3.15	−1.73	0.10	0.00	−0.08	0.28	−0.03	0.02
Mo	0.40	−2.66	−0.64	−0.07	0.34	−0.71	−0.01	0.03	−0.18	−0.36	0.26	−0.05	−0.41	−1.53	−0.16
Ro	0.19	−0.23	−0.21	1.62	0.00	0.83	−1.53	−2.30	3.85	1.57	0.00	1.11	−0.13	−0.13	−0.08
Sm	−0.71	0.13	−1.02	2.90	0.33	0.76	3.77	2.26	0.03	1.40	0.25	1.99	−0.66	0.07	0.29
Fe	−1.19	0.29	−0.79	−1.02	−0.34	0	0	0	0	0	0.26	0.70	0.51	−0.17	−0.49
An	−0.56	0.43	−0.63	1.73	0.11	0	0	0	0	0	−0.08	−1.19	0.41	−0.25	−0.23
De	−0.50	−0.12	1.37	−0.03	0.30	0	0	0	0	0	−0.23	0.02	−0.88	0.07	−0.20
Lo	1.29	0.87	−0.42	−0.50	1.40	0	0	0	0	0	−1.07	0.34	0.27	−0.50	0.53
Am	0.51	−2.18	−0.59	0.01	0.18	0	0	0	0	0	−0.14	−0.01	0.38	1.26	0.21
Se	1.19	0.39	0.22	−0.01	−1.40	0	0	0	0	0	1.07	0.00	−0.14	−0.22	0.49

**Notes* Co = component, Ba = bat, Bl = blood, Bu = butterfly, Ca = cave, Cl = clouds
Fi = fire, Fu = fur, Ma = mask, Mo = mountains, Ro = rocks, Sm = smoke, Fe = fear
An = anger, Dep = depression, Lo = love, Am = ambition, Se = security

Table 5.25 Principal coordinates of dual space

Co*	1	2	3	4	5	11	12	13	14	15
Ba	−1.09	0.26	−0.34	−0.74	−0.24	−0.18	−0.51	−0.22	0.15	0.44
Bl	−1.12	0.63	−1.17	−0.32	−0.17	−0.13	−0.22	−0.75	0.36	0.46
Bu	1.51	0.74	−0.45	−0.51	0.92	0.70	−0.35	−0.29	0.43	−0.62
Ca	−0.47	−0.50	0.89	−0.83	0.13	0.10	−0.57	0.57	−0.29	0.19
Cl	−0.26	0.17	1.53	0.57	0.49	0.37	0.39	0.98	0.10	0.11
Fi	−0.53	0.54	−1.20	1.39	0.23	0.18	0.96	−0.77	0.31	0.22
Fu	1.12	0.12	0.202	0.19	−1.97	−1.50	0.13	0.13	0.07	−0.46
Ma	−0.04	−0.06	0.43	−0.12	0.00	0.00	−0.08	0.28	−0.03	0.02
Mo	0.40	−2.66	−0.64	−0.07	0.34	0.26	−0.05	−0.41	−1.53	−0.16
Ro	0.19	−0.23	−0.21	1.62	0.00	0.00	1.11	−0.13	−0.13	−0.08
Sm	−0.71	0.13	−1.02	2.90	0.33	0.25	1.99	−0.66	0.07	0.29
Fe	−1.19	0.29	−0.79	−1.02	−0.34	0.26	0.70	0.51	−0.17	−0.49
An	−0.56	0.43	−0.63	1.73	0.11	−0.08	−1.19	0.41	−0.25	−0.23
De	−0.50	−0.12	1.37	−0.03	0.30	−0.23	0.02	−0.88	0.07	−0.20
Lo	1.29	0.87	−0.42	−0.50	1.40	−1.07	0.34	0.27	−0.50	0.53
Am	0.51	−2.18	−0.59	0.01	0.18	−0.14	−0.01	0.38	1.26	0.21
Se	1.19	0.39	0.22	−0.01	−1.40	1.07	0.00	−0.14	−0.22	0.49

**Notes* Co = component, Ba = bat, Bl = blood, Bu = butterfly, Ca = cave
Cl = clouds, Fi = fire, Fu = fur, Ma = mask, Mo = mountains, Ro = rocks, Sm = smoke
Fe = fear, An = anger, Dep = depression, Lo = love, Am = ambition, Se = security

5.8.1 Graphical Approach and Further Problems

(a) Pairwise Graph

It has been common practice to use pairwise graphs. Under this practice, it seems that symmetric scaling is a decisive favourite.

As mentioned earlier under the topic of French wisdom, it seems symmetric scaling is more useful than plotting all possible pairs of components in dual space. Then, should we recommend symmetric scaling for pairwise plots? But, how can we discard the information contained in those other extra paired plots than what we use in French plots, or how can we summarize all the information into a manageable form. We need new research on how to extract useful information from all possible pairwise plots in dual space, a topic that requires future research.

One point to note is the fact that as the data table becomes large, we will soon find out that the pairwise comparisons of two components may not simplify the interpretation of the outcome because of too many pairs of components. Then, why should we bother even with pairwise symmetric scaling?

(b) Multidimensional Graphs

The object of our search for the best joint graphical graph is of course to present a multidimensional graph that accommodates all the components. Since our visual ability of grasping multidimensional relations between rows and columns is typically limited to a three-dimensional configuration, we need to develop a program of dynamic interactive visual display. More concretely, we can plot mathematically a multidimensional configuration of data points, and the program should allow the investigator to rotate a multidimensional graph of data configuration to arrive at clusters of variables. How to extract key information through rotations of the multidimensional configuration is a question to be investigated. We can start, for example, with the first three components and let the program allow the initial principal axis configuration to be rotated dynamically under the control of the investigator; one of the three components can be then replaced with another component, and search for the best configuration through rotations of the three axes, and so on. This dynamic trial-and-error process may be carried out interactively in search of the "best" multidimensional configuration, the best in some sense which is a topic for future investigation. Once the parameters of goodness of a configuration are established, the initial subjective rotation can perhaps be translated into the objective process. In other words, this process can eventually be automated to arrive at identifying many sets of clusters of rows and columns in multidimensional space. This learning machine approach may provide a more satisfactory result than the symmetric scaling of pairwise graphs, but we are still a long way to develop such an approach. We may also need a device that generates constantly changing, holographic, dynamic, and interactive configurations of data points, incorporating interactive learning tools, a good research topic for the future.

(c) Dimensionless Approach: Cluster Analysis

To attain a useful tool to interpret the multidimensional outcome of quantification theory, one alternative is to move from the dimensional framework to a non-dimensional framework. What comes to our mind immediately is cluster analysis. Lebart and Mirkin (1993) provided a comparison between the traditional graphical approach and cluster analysis, showing similarities and differences between the two approaches. Their work provides us with some hope that the combined use of these two approaches may lead to a more satisfactory approach than joint graphical display. Furthermore, Nishisato (2014) recommended carrying out *cluster analysis in dual space*, which means cluster analysis of the between-set distances, that is, the distance matrix of rows × columns. Since cluster analysis of dual space is a quite different approach from multidimensional graphical display, Clavel and Nishisato will discuss this approach fully in the next chapter.

5.8.2 Within-Set Distance in Dual Space

When the space is expanded from contingency space to dual space, we should realize the within-set distance in contingency space is typically smaller than the correspondence distance in dual space. Note that the distance between two points in smaller dimensional space cannot be larger than the distance between the same two points in larger dimensional space. The traditional within-set distance (see Greenacre 1984; Nishisato and Clavel 2010) was defined for a contingency table, and cannot be used for analysis in dual space.

To rectify this difference, Nishisato (2019) presented a formula for the within-set distance in dual space, calculated from the contingency-table analysis. Assuming that the distribution of points is centred for each dimension, we again consider half the discrepancy angle to place data points away from the axis. Thus, the coordinates $\rho_k y_{ki}$ and $\rho_k x_{kj}$ should be replaced by

- $\rho_k y_{ki} cos(\frac{\theta_k}{2})$ and $\rho_k x_{kj} cos(\frac{\theta_k}{2})$, respectively.
- These can be further rewritten as $\frac{\rho_k^2}{2} y_{ki}$ and $\frac{\rho_k^2}{2} x_{kj}$, respectively.

Notice that the above formula provides a way to calculate the correct within-set distance, that is, the within-set distances in dual space from the results of the contingency-table analysis. A much simpler way than this is to use the results of the response-pattern table, for we then need to calculate the within-set distance directly from the set of principal coordinates.

Let us now look at cluster analysis as an alternative to joint graphical display in the next chapter.

References

Garmize, L. M., & Rychlak, J. F. (1964). Role-play validation of a socio-cultural theory of symbolism. *Journal of Consulting Psychology, 28,* 107–115.

Greenacre, M. J. (1984). *Theory and applications of correspondence analysis*. London: Academic.

Lebart, L., Morineau, A., & Tabard, N. (1977). *Techniques de la Description Statistique: Méthodes et Logiciels pour l'Analyse des Grands Tableaux*. Paris: Dunod.

Lebart, L., & Mirkin, B.D. (1993). Correspondence analysis and classification. *Multivariate Analysis: Future Directions* 2, 341–357.

Nishisato, S. (1980). *Analysis of categorical data: Dual scaling and its applications*. Toronto: The University of Toronto Press.

Nishisato, S. (2014). Structural representation of categorical data and cluster analysis through filters. In Gaul, W., Geyer-Schultz, A., Baba, Y. and Okada, A. (Eds.)., *German- Japanese Interchange of Data Analysis Results*. Springer, 81–90.

Nishisato, S. (2019). Reminiscence: Quantification theory and graphs. *Theory and Applications of Data Analysis*, *8*, 47–57 (in Japanese).

Nishisato, S., & Clavel, J. G. (2010). Total information analysis: Comprehensive dual scaling. *Behaviormetrika, 37,* 15–32.

Chapter 6
Clustering as an Alternative

We have now a huge amount of cluster analysis literature. However, most of them do not handle the following situations that we are dealing with:

1. We are primarily interested in the row-column relations of the contingency tables, thus cluster analysis of the between-set distance matrices (i.e., row-by-column distance matrices). Unless the number of rows is equal to that of columns, this distance matrix is *rectangular*.
2. Even if the between-set matrix is square, it is not symmetric (*a typical condition imposed on hierarchical clustering and partitioning*).
3. In order to maintain the additivity of quantification subspaces, we must use the squared distances, as we will show it shortly.
4. To pursue cluster structure of row-column relations, we will consider two matrices of squared distance, namely the between-set (rows and columns) distance matrix and super-distance matrix in dual space.
5. Since our use of cluster analysis lies in the question of how appropriate cluster analysis is to describe the row-column relations, we will explore clusters in contingency space and dual space. Keep in mind that contingency space ignores the contributions of supplementary space, the space associated with eigenvalues smaller than 0.5 of the response-pattern matrix in dual space.

These five aspects of our task make the current work quite unique. As for the clustering algorithms for symmetric distance matrices, we have investigated currently available algorithms of (1) hierarchical clustering, (2) partitioning clustering, and (3) bi-clustering. Our main conclusion is that hierarchical clustering provides the most "reasonable" outcome, followed by partitioning such as the k-means clustering. To our dismay, bi-clustering did not show much promise from our point of view as users of these methods. From this preliminary investigation, we have decided to use hierarchical and partitioning methods as alternative clustering approaches.

The main obstacle for these clustering methods from our point of view is that these methods are based on analysis of symmetric input matrices, and our between-set matrix, even when it is square, is not symmetric. This prevents us from using

S. Nishisato et al., *Modern Quantification Theory*,
Behaviormetrics: Quantitative Approaches to Human Behavior 8,
https://doi.org/10.1007/978-981-16-2470-4_6

these clustering methods. To overcome this practical problem, we have developed a method to transform the input matrix so as to make the input matrix amenable to clustering by the popular method for the square symmetric matrix. Our proposed method is called *Universal Transform for Clustering* (UTC). Note that this is not a method of cluster analysis, but a method to transform the input matrix so as to make it amenable to any existing methods of cluster analysis. UTC will prove to be handy when we consider clustering in the between-set distance matrix.

We will make several conclusions on the use of cluster analysis as an alternative to joint graphical display.

6.1 Decomposition of Input Data

Let us first look at a variety of input matrices for cluster analysis, starting with the total super-distance matrix (i.e., the (rows, columns)-by-(rows, columns) matrix), partitioned into (dual space + residual space), or (contingency space + supplementary space + residual space) as discussed in Chap. 5. Although our main objective is to look at clusters embedded in between-set distance matrices in dual space, we should at least know how distance matrices in different quantification space would look like, using two numerical examples, namely the Rorschach data (Garmize and Rycklak 1964) and barley data (Stebbins 1950).

6.1.1 Rorschach Data

We have used this data set in previous chapters. As before, we will adopt the 11×6 matrix, instead of the original 16×6 contingency table. In Chap. 5, this matrix was first transformed into the response-pattern table, and we obtained the 17×15 table of principal coordinates, where the rows consist of 11 Rorschach inkblots and 6 moods, and columns are 15 components (Table 5.24). Let us indicate this matrix by $\mathbf{Z}$, with elements z_{ij}.

6.1.2 Barley Data

Stebbins (1950) provided a data set which can be described as a demonstration of a mini-theory of evolution: six agricultural experimental stations were chosen, and six varieties of barley were planted at those stations; during harvest time, 500 seeds were randomly chosen in each station and classified into the six varieties; next season, these seeds were planted at the six stations; during the harvest, 500 seeds were randomly chosen and sorted into the six varieties in each station. This process was repeated a number of times to show that the varieties of barley will gradually indicate which

Table 6.1 After many years of barley varieties at different locations

	Arlington (Arl)	Ithaca (Ith)	St. Paul (StP)	Moccasin (Moc)	Moro (Mor)	Davis (Dav)
Coast-Trebi (Co)	446	57	83	87	6	362
Hanchen (Ha)	4	34	305	19	4	34
White Smyrna (Wh)	4	0	4	241	489	65
Manchuria (Ma)	1	343	2	21	0	0
Gatemi (Ga)	13	9	15	58	0	1
Meloy (Me)	4	0	0	4	0	27

Table 6.2 Basic statistics for the response patterns of barley data

Component	Eigenvalue	SingValue	Delta	CumDelta
1	0.9287	0.9637	18.5746	18.57461
2	0.8976	0.9474	17.9528	36.52736
3	0.8465	0.9201	16.9299	53.45730
4	0.6210	0.7881	12.4207	65.87799
5	0.5690	0.7543	11.3793	77.25726
6	0.4310	0.6565	8.6207	85.87799
7	0.3790	0.6156	7.5793	93.45730
8	0.1535	0.3918	3.0701	96.52736
9	0.1024	0.3199	2.0472	98.57461
10	0.0713	0.2670	1.4254	100.00000

stations are the best for the six varieties of barley. This data set, presented in Table 6.1, was first used by Nishisato (1994), and then used in cluster-analytic studies by Clavel and Nishisato (2008, 2012, 2020), Nishisato and Clavel (2008) and Nishisato (2012, 2014).

Unlike the Rorschach data, the barley data set is an example in which the number of rows is equal to the number of columns. This means that there is no residual space involved, leaving the space structure simple, namely total space is the same as dual space.

As was the case with Rorschach data, we quantify the data, first transforming it to the response-pattern matrix and then arriving at the matrix of Euclidean coordinates (i.e., principal coordinates) **Z**. The basic statistics from dual scaling are presented in Table 6.2. Notice that there are no eigenvalues equal to 0.5, indicating no residual space involved in this data set. The sum of the 10 eigenvalues is equal to 5. The sum

Table 6.3 Space partitions for Rorschach data

Space	Cont	Dual	Supp	Resid	Total
$\sum \rho^2$	3.61	5.00	1.39	2.50	7.50
Percentage (%)	51	71	20	29	100

of the eigenvalues of the contingency space is 3.863, which is 77.25% of dual space. Thus, dual space contains extra 22.75% of information.

To maintain the additive structure of space decomposition, we must adopt the *squared distances* as input for cluster analysis, namely the squared distance between variables s and t of $\mathbf{Z}$. In our example of Rorschach data, we use the 17×15 matrix, $\mathbf{Z}$, of principal coordinates, and the elements of the matrix of the squared distance can be calculated by

$$d_{st}^2 = \frac{\sum_j (z_{sj} - z_{tj})^2}{n + m - 2}. \tag{6.1}$$

For the Rorschach data, the total space involves 15 components (see Chap. 5), of which the first five components belong to contingency space, components 11–15 to supplementary space. Thus, dual space consists of components 1–5 and 11–15. Residual space consists of components 6–10. The contribution of each space in terms of the percentages of the corresponding eigenvalue is presented on Table 6.3.

Let us indicate the $(m + n) \times (m + m)$ matrix of elements d_{st} with several partitions, namely $\mathbf{D^2}_{total}$ for total space, $\mathbf{D^2}_{contingency}$ for contingency space, $\mathbf{D^2}_{dual}$ for dual space, $\mathbf{D^2}_{supplementary}$ for supplementary space, and $\mathbf{D^2}_{residual}$ for residual space.

The reason why we use *squared* distances is solely to attain the following additive relations. (Notice that *these additive relations do not exist for distances unless they are squared!*):

$$\mathbf{D^2}_{total} = \mathbf{D^2}_{contingency} + \mathbf{D^2}_{supplementary} + \mathbf{D^2}_{residual} \tag{6.2}$$

$$\mathbf{D^2}_{total} = \mathbf{D^2}_{dual} + \mathbf{D^2}_{residual}. \tag{6.3}$$

Before moving on, let us look at squared distance matrices in the entire quantification space for the Rorschach example. The squared super-distance matrix in total space, $\mathbf{D^2}_{total}$, is as shown in Table 6.4. This can be partitioned into $\mathbf{D^2}_{dual}$ (Table 6.5), $\mathbf{D^2}_{residual}$ (Table 6.6), $\mathbf{D^2}_{contingency}$ (Table 6.7), and $\mathbf{D^2}_{supplementary}$ (Table 6.8).

Note, for example, the squared total distance between the mood fear and the symbol blood, 1.2264, can be decomposed in two ways: as the sum of those in dual space and residual space (i.e., $1.2264 = 0.3280 + 0.8985$) or as the sum of those in contingency space (0.0562), supplementary space (0.2717), and residual subspace (0.8985). Observe the additivity of subspace distances. One can infer from the amount of information loss by looking at contingency space as opposed to dual space. If we calculate this statistic for individual variables, we obtain, for example, *ambition* 7.1% and *anger* 3.8%.

Already suggested on Table 6.3, the *relative contributions* of partitioned total space can also be assessed by the relative magnitudes of the distance on the corresponding tables.

Let us now look at our second example, barley data. The matrix of principal coordinates is 12×12. Table 6.9 shows the relative sizes of space decomposition. Since the number of rows is equal to that of columns, there is no residual space. Consequently, total space is the same with dual space. The relative size of contingency space to dual space is 77.2%, meaning that we miss 22.8% of information by limiting quantification space to contingency space.

We have shown the partitions of total space, using two sets of data. The purpose of this demonstration is (1) to show how important it is to collect data using the same numbers of rows and columns (i.e., no residual space): Examine the contributions of distances in residual space, associated with Rorschach data, which can be regarded as contamination to the row-column relations, and (2) how much information we miss by looking at only contingency space, rather than dual space. Since the second point depends on data sets, we will leave to the readers the examination of the distance matrices in supplementary space and residual space (Rorschach data). One should look for relatively large distances in supplementary space which the analysis of contingency space would miss, leading to the distortion of the final configurations or cluster formations. One should particularly pay attention to the residual space, which does not involve the row-column relations, thus is irrelevant to our purpose of pursuing row-column relations. The distance in residual space can often be substantial. This tells us two important lessons, one that distances in residual space may adversely influence the outcome of the analysis and the other that equating the number of rows and that of columns, so long as it is possible, is very important.

6.2 Partitions of Super-Distance Matrix

Let us define the squared super-distance matrix $\mathbf{D}^2$, consisting of squared within-row $\mathbf{D}_r^2$, squared within-column $\mathbf{D}_c^2$, squared between-row-columns $\mathbf{D}_{rc}^2$, and squared between-column-rows $\mathbf{D}_{cr}^2$ distance matrices, arranged as

$$\begin{bmatrix} \mathbf{D}_r^2 & \mathbf{D}_{rc}^2 \\ \mathbf{D}_{cr}^2 & \mathbf{D}_c^2 \end{bmatrix}.$$

In terms of our Rorschach example, $\mathbf{D}_r^2$ is 11×11, $\mathbf{D}_{rc}^2$ is 11×6, $\mathbf{D}_{cr}^2$ is 6×11, and $\mathbf{D}_c^2$ is 6×6. In the Barley example, $\mathbf{D}_r^2$ is 6×6, so are $\mathbf{D}_{rc}^2$, $\mathbf{D}_{cr}^2$, and $\mathbf{D}_c^2$ since $n = m = 6$.

Up to this point, we have not defined the between-set distance matrix and the within-set distance matrix. As may be clear from the above descriptions, we call $\mathbf{D}_r^2$

Table 6.4 Total space (Rorschach data)

	Bat	Blood	Butterfly	Cave	Clouds	Fire	Fur	Mask	Mountains	Rocks	Smoke	Fear	Anger	Depression	Love	Ambition	Security
Bat	0	1.5924	0.7641	1.1701	0.7642	1.5257	0.9257	1.5924	1.14	2.3985	2.8591	0.3464	0.6451	0.4608	0.8291	0.85	0.6354
Blood	1.5924	0	1.7051	2.1111	1.7051	2.4667	1.8667	2.5333	2.0809	3.3394	3.8	1.2264	1.4654	1.4829	1.7272	1.8227	1.6404
Butterfly	0.7641	1.7051	0	1.2829	0.8769	1.6385	1.0385	1.7052	1.2527	2.5112	2.9719	0.8004	0.8511	0.7058	0.4385	0.8876	0.5535
Cave	1.1701	2.1111	1.2829	0	1.2829	2.0444	1.4445	2.1111	1.6587	2.9172	3.3778	1.0187	1.2915	0.8547	1.3242	1.2358	1.1628
Clouds	0.7642	1.7051	0.8769	1.2829	0	1.6385	1.0385	1.7051	1.2527	2.5112	2.9718	0.7725	0.7308	0.3957	0.8769	0.9732	0.726
Fire	1.5257	2.4667	1.6385	2.0444	1.6385	0	1.8	2.4667	2.0143	3.2728	3.7334	1.3714	1.2236	1.4488	1.6091	1.6976	1.5344
Fur	0.9257	1.8667	1.0385	1.4445	1.0385	1.8	0	1.8667	1.4143	2.6728	3.1334	0.9619	0.9765	0.8195	0.9819	1.0098	0.5607
Mask	1.5924	2.5333	1.7052	2.1111	1.7051	2.4667	1.8667	0	2.081	3.3394	3.8	1.5079	1.6144	1.3594	1.6697	1.6992	1.5158
Mountains	1.14	2.0809	1.2527	1.6587	1.2527	2.0143	1.4143	2.081	0	2.8871	3.3477	1.1245	1.2294	1.0129	1.2954	0.6554	1.1346
Rocks	2.3985	3.3394	2.5112	2.9172	2.5112	3.2728	2.6728	3.3394	2.8871	0	4.6061	2.4347	2.1947	2.2497	2.4967	2.4267	2.3106
Smoke	2.8591	3.8	2.9719	3.3778	2.9718	3.7334	3.1334	3.8	3.3477	4.6061	0	2.8149	2.3844	2.7496	3.0515	2.9659	2.9071
Fear	0.3464	1.2264	0.8004	1.0187	0.7725	1.3714	0.9619	1.5079	1.1245	2.4347	2.8149	0	0.809	0.64	0.8801	0.918	0.7357
Anger	0.6451	1.4654	0.8511	1.2915	0.7308	1.2236	0.9765	1.6144	1.2294	2.1947	2.3844	0.809	0	0.7251	0.9653	1.0032	0.8208
Depression	0.4608	1.4829	0.7058	0.8547	0.3957	1.4488	0.8195	1.3594	1.0129	2.2497	2.7496	0.64	0.7251	0	0.7962	0.8342	0.6518
Love	0.8291	1.7272	0.4385	1.3242	0.8769	1.6091	0.9819	1.6697	1.2954	2.4967	3.0515	0.8801	0.9653	0.7962	0	1.0743	0.892
Ambition	0.85	1.8227	0.8876	1.2358	0.9732	1.6976	1.0098	1.6992	0.6554	2.4267	2.9659	0.918	1.0032	0.8342	1.0743	0	0.9298
Security	0.6354	1.6404	0.5535	1.1628	0.726	1.5344	0.5607	1.5158	1.1346	2.3106	2.9071	0.7357	0.8208	0.6518	0.892	0.9298	0

Table 6.5 Dual space (Rorschach data)

	Bat	Blood	Butterfly	Cave	Clouds	Fire	Fur	Mask	Mountains	Rocks	Smoke	Fear	Anger	Depression	Love	Ambition	Security
Bat	0	0.1258	0.6904	0.2468	0.6027	0.5795	0.7264	0.1818	1.0238	0.6868	1.3929	0.2418	0.5404	0.3562	0.7244	0.7454	0.5308
Blood	0.1258	0	0.7261	0.5893	0.9055	0.3375	1.1139	0.4051	1.2094	0.7092	1.0959	0.328	0.5669	0.5845	0.8288	0.9243	0.7419
Butterfly	0.6904	0.7261	0	0.6912	0.7829	0.7952	0.9988	0.4171	1.1966	0.774	1.6431	0.7916	0.8424	0.697	0.4297	0.8788	0.5447
Cave	0.2468	0.5893	0.6912	0	0.2904	1.0032	0.9155	0.1083	0.7572	0.7731	1.7665	0.4349	0.7077	0.2709	0.7404	0.652	0.579
Clouds	0.6027	0.9055	0.7829	0.2904	0	0.8064	0.9559	0.1894	1.2433	0.458	1.177	0.6889	0.6471	0.312	0.7933	0.8895	0.6423
Fire	0.5795	0.3375	0.7952	1.0032	0.8064	0	1.1008	0.538	1.2329	0.2153	0.2476	0.6289	0.481	0.7062	0.8665	0.955	0.7919
Fur	0.7264	1.1139	0.9988	0.9155	0.9559	1.1008	0	0.5607	1.3569	0.669	1.646	0.9386	0.9532	0.7962	0.9586	0.9865	0.5373
Mask	0.1818	0.4051	0.4171	0.1083	0.1894	0.538	0.5607	0	0.7375	0.3507	1.1436	0.3278	0.4343	0.1793	0.4896	0.5191	0.3358
Mountains	1.0238	1.2094	1.1966	0.7572	1.2433	1.2329	1.3569	0.7375	0	0.8582	1.6875	1.0776	1.1825	0.966	1.2485	0.6085	1.0877
Rocks	0.6868	0.7092	0.774	0.7731	0.458	0.2153	0.669	0.3507	0.8582	0	0.3121	0.7047	0.4648	0.5198	0.7668	0.6967	0.5806
Smoke	1.3929	1.0959	1.6431	1.7665	1.177	0.2476	1.646	1.1436	1.6875	0.3121	0	1.3309	0.9004	1.2656	1.5675	1.4819	1.4231
Fear	0.2418	0.328	0.7916	0.4349	0.6889	0.6289	0.9386	0.3278	1.0776	0.7047	1.3309	0	0.809	0.64	0.8801	0.918	0.7357
Anger	0.5404	0.5669	0.8424	0.7077	0.6471	0.481	0.9532	0.4343	1.1825	0.4648	0.9004	0.809	0	0.7251	0.9653	1.0032	0.8208
Depression	0.3562	0.5845	0.697	0.2709	0.312	0.7062	0.7962	0.1793	0.966	0.5198	1.2656	0.64	0.7251	0	0.7962	0.8342	0.6518
Love	0.7244	0.8288	0.4297	0.7404	0.7933	0.8665	0.9586	0.4896	1.2485	0.7668	1.5675	0.8801	0.9653	0.7962	0	1.0743	0.892
Ambition	0.7454	0.9243	0.8788	0.652	0.8895	0.955	0.9865	0.5191	0.6085	0.6967	1.4819	0.918	1.0032	0.8342	1.0743	0	0.9298
Security	0.5308	0.7419	0.5447	0.579	0.6423	0.7919	0.5373	0.3358	1.0877	0.5806	1.4231	0.7357	0.8208	0.6518	0.892	0.9298	0

Table 6.6 Residual space (Rorschach data)

	Bat	Blood	Butterfly	Cave	Clouds	Fire	Fur	Mask	Mountains	Rocks	Smoke	Fear	Anger	Depression	Love	Ambition	Security
Bat	0	1.4665	0.0738	0.9234	0.1614	0.9462	0.1993	1.4105	0.1162	1.7117	1.4662	0.1046	0.1046	0.1046	0.1046	0.1046	0.1046
Blood	1.4665	0	0.979	1.5217	0.7997	2.1291	0.7527	2.1282	0.8715	2.6303	2.7041	0.8985	0.8985	0.8985	0.8985	0.8985	0.8985
Butterfly	0.0738	0.979	0	0.5916	0.094	0.8432	0.0396	1.288	0.0561	1.7372	1.3288	0.0088	0.0088	0.0088	0.0088	0.0088	0.0088
Cave	0.9234	1.5217	0.5916	0	0.9925	1.0412	0.529	2.0029	0.9015	2.144	1.6114	0.5838	0.5838	0.5838	0.5838	0.5838	0.5838
Clouds	0.1614	0.7997	0.094	0.9925	0	0.8321	0.0826	1.5158	0.0094	2.0532	1.7948	0.0837	0.0837	0.0837	0.0837	0.0837	0.0837
Fire	0.9462	2.1291	0.8432	1.0412	0.8321	0	0.6992	1.9287	0.7814	3.0574	3.4858	0.7426	0.7426	0.7426	0.7426	0.7426	0.7426
Fur	0.1993	0.7527	0.0396	0.529	0.0826	0.6992	0	1.3061	0.0574	2.0038	1.4874	0.0233	0.0233	0.0233	0.0233	0.0233	0.0233
Mask	1.4105	2.1282	1.288	2.0029	1.5158	1.9287	1.3061	0	1.3435	2.9887	2.6564	1.1801	1.1801	1.1801	1.1801	1.1801	1.1801
Mountains	0.1162	0.8715	0.0561	0.9015	0.0094	0.7814	0.0574	1.3435	0	2.0289	1.6602	0.0469	0.0469	0.0469	0.0469	0.0469	0.0469
Rocks	1.7117	2.6303	1.7372	2.144	2.0532	3.0574	2.0038	2.9887	2.0289	0	4.294	1.7299	1.7299	1.7299	1.7299	1.7299	1.7299
Smoke	1.4662	2.7041	1.3288	1.6114	1.7948	3.4858	1.4874	2.6564	1.6602	4.294	0	1.484	1.484	1.484	1.484	1.484	1.484
Fear	0.1046	0.8985	0.0088	0.5838	0.0837	0.7426	0.0233	1.1801	0.0469	1.7299	1.484	0	0	0	0	0	0
Anger	0.1046	0.8985	0.0088	0.5838	0.0837	0.7426	0.0233	1.1801	0.0469	1.7299	1.484	0	0	0	0	0	0
Depression	0.1046	0.8985	0.0088	0.5838	0.0837	0.7426	0.0233	1.1801	0.0469	1.7299	1.484	0	0	0	0	0	0
Love	0.1046	0.8985	0.0088	0.5838	0.0837	0.7426	0.0233	1.1801	0.0469	1.7299	1.484	0	0	0	0	0	0
Ambition	0.1046	0.8985	0.0088	0.5838	0.0837	0.7426	0.0233	1.1801	0.0469	1.7299	1.484	0	0	0	0	0	0
Security	0.1046	0.8985	0.0088	0.5838	0.0837	0.7426	0.0233	1.1801	0.0469	1.7299	1.484	0	0	0	0	0	0

Table 6.7 Contingency space (Rorschach data)

	Bat	Blood	Butterfly	Cave	Clouds	Fire	Fur	Mask	Mountains	Rocks	Smoke	Fear	Anger	Depression	Love	Ambition	Security
Bat	0	0.0879	0.5447	0.1763	0.4213	0.3972	0.5311	0.1351	0.7743	0.4836	0.9479	0.0275	0.4397	0.2799	0.5937	0.6203	0.387
Blood	0.0879	0	0.5882	0.4254	0.6447	0.2338	0.8036	0.3004	0.9238	0.51	0.7478	0.0562	0.3462	0.5303	0.5999	0.7633	0.6225
Butterfly	0.5447	0.5882	0	0.537	0.587	0.5984	0.6444	0.32	0.9012	0.5512	1.1858	0.6337	0.6799	0.5854	0.02	0.6999	0.4124
Cave	0.1763	0.4254	0.537	0	0.2013	0.7002	0.6314	0.0774	0.5614	0.5329	1.213	0.2822	0.6614	0.072	0.5625	0.4568	0.4703
Clouds	0.4213	0.6447	0.587	0.2013	0	0.5694	0.6512	0.1322	0.9136	0.3212	0.8196	0.6334	0.4269	0.0367	0.5825	0.7445	0.5024
Fire	0.3972	0.2338	0.5984	0.7002	0.5694	0	0.7649	0.3781	0.9158	0.1583	0.1697	0.4545	0.0347	0.6158	0.6056	0.7317	0.6373
Fur	0.5311	0.8036	0.6444	0.6314	0.6512	0.7649	0	0.3773	0.9521	0.4515	1.1456	0.7222	0.6787	0.6137	0.8461	0.725	0.0316
Mask	0.1351	0.3004	0.32	0.0774	0.1322	0.3781	0.3773	0	0.5497	0.2398	0.7893	0.2584	0.3412	0.0781	0.3615	0.399	0.2422
Mountains	0.7743	0.9238	0.9012	0.5614	0.9136	0.9158	0.9521	0.5497	0	0.6198	1.213	0.8414	0.9311	0.7599	0.9864	0.0186	0.8949
Rocks	0.4836	0.51	0.5512	0.5329	0.3212	0.1583	0.4515	0.2398	0.6198	0	0.2251	0.6483	0.0818	0.3929	0.6063	0.4543	0.3954
Smoke	0.9479	0.7478	1.1858	1.213	0.8196	0.1697	1.1456	0.7893	1.213	0.2251	0	1.0764	0.111	0.9706	1.1891	1.0366	1.088
Fear	0.0275	0.0562	0.6337	0.2822	0.6334	0.4545	0.7222	0.2584	0.8414	0.6483	1.0764	0	0.5536	0.4526	0.6731	0.7009	0.555
Anger	0.4397	0.3462	0.6799	0.6614	0.4269	0.0347	0.6787	0.3412	0.9311	0.0818	0.111	0.5536	0	0.5051	0.697	0.7411	0.5842
Depression	0.2799	0.5303	0.5854	0.072	0.0367	0.6158	0.6137	0.0781	0.7599	0.3929	0.9706	0.4526	0.5051	0	0.5951	0.6181	0.4654
Love	0.5937	0.5999	0.02	0.5625	0.5825	0.6056	0.8461	0.3615	0.9864	0.6063	1.1891	0.6731	0.697	0.5951	0	0.7902	0.5719
Ambition	0.6203	0.7633	0.6999	0.4568	0.7445	0.7317	0.725	0.399	0.0186	0.4543	1.0366	0.7009	0.7411	0.6181	0.7902	0	0.6669
Security	0.387	0.6225	0.4124	0.4703	0.5024	0.6373	0.0316	0.2422	0.8949	0.3954	1.088	0.555	0.5842	0.4654	0.5719	0.6669	0

Table 6.8 Supplementary space (Rorschach data)

	Bat	Blood	Butterfly	Cave	Clouds	Fire	Fur	Mask	Mountains	Rocks	Smoke	Fear	Anger	Depression	Love	Ambition	Security
Bat	0	0.0379	0.1457	0.0704	0.1815	0.1824	0.1953	0.0468	0.2495	0.2032	0.445	0.2143	0.1007	0.0763	0.1308	0.1251	0.1438
Blood	0.0379	0	0.1379	0.164	0.2607	0.1037	0.3103	0.1047	0.2856	0.1992	0.3481	0.2717	0.2207	0.0542	0.2289	0.161	0.1195
Butterfly	0.1457	0.1379	0	0.1543	0.1958	0.1969	0.3544	0.0971	0.2954	0.2228	0.4573	0.1579	0.1624	0.1116	0.4097	0.1789	0.1323
Cave	0.0704	0.164	0.1543	0	0.0891	0.303	0.2841	0.0308	0.1958	0.2402	0.5535	0.1527	0.0463	0.199	0.1779	0.1951	0.1087
Clouds	0.1815	0.2607	0.1958	0.0891	0	0.2371	0.3047	0.0572	0.3298	0.1368	0.3574	0.0555	0.2202	0.2753	0.2107	0.145	0.1399
Fire	0.1824	0.1037	0.1969	0.303	0.2371	0	0.3359	0.1599	0.3171	0.057	0.0779	0.1744	0.4463	0.0905	0.2609	0.2233	0.1545
Fur	0.1953	0.3103	0.3544	0.2841	0.3047	0.3359	0	0.1834	0.4049	0.2175	0.5004	0.2164	0.2745	0.1825	0.1125	0.2614	0.5057
Mask	0.0468	0.1047	0.0971	0.0308	0.0572	0.1599	0.1834	0	0.1878	0.1109	0.3543	0.0694	0.0931	0.1012	0.1281	0.1201	0.0936
Mountains	0.2495	0.2856	0.2954	0.1958	0.3298	0.3171	0.4049	0.1878	0	0.2384	0.4745	0.2362	0.2514	0.2061	0.2621	0.59	0.1929
Rocks	0.2032	0.1992	0.2228	0.2402	0.1368	0.057	0.2175	0.1109	0.2384	0	0.087	0.0564	0.383	0.1269	0.1604	0.2424	0.1852
Smoke	0.445	0.3481	0.4573	0.5535	0.3574	0.0779	0.5004	0.3543	0.4745	0.087	0	0.2545	0.7894	0.2951	0.3785	0.4453	0.3351
Fear	0.2143	0.2717	0.1579	0.1527	0.0555	0.1744	0.2164	0.0694	0.2362	0.0564	0.2545	0	0.2554	0.1874	0.207	0.2171	0.1806
Anger	0.1007	0.2207	0.1624	0.0463	0.2202	0.4463	0.2745	0.0931	0.2514	0.383	0.7894	0.2554	0	0.22	0.2682	0.2621	0.2366
Depression	0.0763	0.0542	0.1116	0.199	0.2753	0.0905	0.1825	0.1012	0.2061	0.1269	0.2951	0.1874	0.22	0	0.2012	0.216	0.1864
Love	0.1308	0.2289	0.4097	0.1779	0.2107	0.2609	0.1125	0.1281	0.2621	0.1604	0.3785	0.207	0.2682	0.2012	0	0.2841	0.3201
Ambition	0.1251	0.161	0.1789	0.1951	0.145	0.2233	0.2614	0.1201	0.59	0.2424	0.4453	0.2171	0.2621	0.216	0.2841	0	0.263
Security	0.1438	0.1195	0.1323	0.1087	0.1399	0.1545	0.5057	0.0936	0.1929	0.1852	0.3351	0.1806	0.2366	0.1864	0.3201	0.263	0

Table 6.9 Space partitions of barley data

Space	Cont	Dual	Supp	Resid	Total
$\sum \rho^2$	3.86	5.00	1.34	0.00	5.00
Percentage	77.2%	100%	22.8%	0	100%

and $\mathbf{D}^2_{\mathbf{c}}$ *within-set* squared distance matrices, and $\mathbf{D}^2_{\mathbf{rc}}$ and $\mathbf{D}^2_{\mathbf{cr}}$ *between-set* squared distance matrices. Within-set matrices are always square, while between-set matrices are rectangular unless the number of rows is equal to that of columns. Usually, our main interest lies in the between-set relationships.

In this chapter, we are mainly interested in the analysis of the between-set distance matrix in dual space. However, because the French plot is a graphical method of rows and columns in contingency space, we will investigate cluster analysis of both dual space and contingency space. At the same time, we must investigate clustering of the super-distance matrix, which is always square and symmetric, for most of the currently popular clustering algorithms handle only symmetric square matrices.

6.3 Outlines of Cluster Analysis

Our main tasks are threefold: (1) to compare cluster structures of distance relations in dual space and contingency space, using two data sets, Rorschach data which involve residual space and barley data which do not; (2) to look at analysis of the super-distance matrix and the corresponding between-set distance matrix. This is primarily to satisfy our interest if the analysis of the between-set data is sufficient to capture the cluster structure of the data; (3) to investigate these tasks by using different approaches to cluster analysis, hierarchical clustering, and partitioning. We know that these different approaches can yield quite different cluster structures.

As mentioned earlier, the between-set distance matrix is either rectangular or square and even when it is square it is not symmetric. For us to use popularly available clustering algorithms for non-symmetric square matrices and rectangular matrices, we will propose a framework, called *clustering of modified data matrices* (CMDM). This framework will allow us to cluster any sub-matrix of the square super-distance matrix, rectangular or square, and symmetric or non-symmetric.

6.3.1 Universal Transform for Clustering (UTC)

Let us start with some background for this universal scheme. Since cluster analysis is a technique to search for variables which are closely positioned in multidimensional space, extremely large distances are ignored in the process of finding clusters. Thus,

Nishisato (2012) proposed a method where relatively large distances are ignored from clustering through *discarding* them from the data set. Later he called this *cluster analysis with flexible filters* (Nishisato 2014). His method starts with discarding all distances greater than the p-percentile point, leaving only a small number of elements for clustering. The smaller the percentile point, the larger the number of distances discarded, leaving a fewer distances for clustering. Its applications are reported in Nishisato (2012, 2014) and Clavel and Nishisato (2020).

Another relevant approach to the same problem is forced classification (Nishisato 1984), based on the concept of projection: Multiplying a subset of data by a large number, prior to quantification, has the effect of projecting the data onto a subset of data (see also Nishisato and Baba 1999). Other applications of projection operators to the quantification problem can be found (see, for example, Nishisato and Lawrence 1989).

The idea of projection onto a subspace can also be extended to ignoring a part of data. To make this strategy useful for currently popular methods of clustering, Clavel came up with an idea of *increasing* the values of those distance measures so that they would be ignored from clustering. This idea is to make the distances large enough to be ignored, in contrast to the clustering through the filter which discards them from the data set. For this book, our procedure is called "the universal transform for clustering" (UTC). By not creating any vacant cells in the super-distance matrix, our UTC transforms the distance matrix in such a way that we can use currently available algorithms (*e.g.*, hierarchical clustering and partitioning).

To explain UTC, let us rewrite the squared super-distance matrix (note: we call the squared distance simply as distance from now on) as follows:

$$\begin{bmatrix} \mathbf{D}_{\mathbf{r}}^2 + a\mathbf{1}\mathbf{1}' & \mathbf{D}_{\mathbf{rc}}^2 + b\mathbf{1}\mathbf{1}' \\ \mathbf{D}_{\mathbf{cr}}^2 + c\mathbf{1}\mathbf{1}' & \mathbf{D}_{\mathbf{c}}^2 + d\mathbf{1}\mathbf{1}' \end{bmatrix}.$$

Suppose, for example, that we have a well-known program for clustering distance measures in a square symmetric matrix. Then, we can carry out clustering of, for example, the between-set distance matrix by setting $a = c = d = Q$, $b = 1$ where Q is a number which makes the smallest element in the other ignored matrices larger than the largest element in the between-set matrix. This clustering would approximate the cluster analysis of $\mathbf{D}_{\mathbf{rc}}^2$ because the large additive constant to the elements of the other three sub-matrices takes them away from the key role in forming clusters. The word "transform" was used because this scheme is not a method of cluster analysis but a method of data transformation. With UTC, clustering of the rectangular matrix or any sub-matrix, symmetric or not, can be analysed through hierarchical or partitioning clustering of the transformed super-distance matrix. This is a simple idea of analysing modified data matrices to suit the purpose of analysis. However, unlike Nishisato's forced classification, based on the logic of mathematical convergence, there is no guarantee that UTC clusters are identical to those obtained by proper algorithms. Note that if one can develop a method of cluster analysis for a rectangular matrix or a non-symmetric square matrix, we must examine how good our UTC works. This is one task we must leave for future research. The current chapter demonstrates that our UTC works well.

6.4 Clustering of Super-Distance Matrix

6.4.1 Hierarchical Cluster Analysis: Rorschach Data

The input matrix is always square and symmetric. For hierarchical clustering of the super-distance matrix, we use the program `hlclust` in `R` based on Ward's method Ward (1963). The first example of Rorschach data yields dendrograms as shown in Figs. 6.1 and 6.2. It is quite encouraging to know that if we consider six clusters, there is no difference between the two, namely $(Anger; Smoke, Fire, Rocks)$, $(Ambition; Mountains)$, $(Depression; Clouds, Cave, Mask)$, $(Fear; Bat, Blood)$, $(Love; Butterfly)$, and $(Security; Fur)$. Subjectively speaking, these six clusters of moods and inkblots make sense. If we further examine the clustering by increasing the number of clusters, we see a number of differences between those clusters in dual space and those in contingency space. From the practical point of view, we wonder if these small differences would matter. Here, our conclusion is that contingency space provides a good approximation to the cluster structure of dual space.

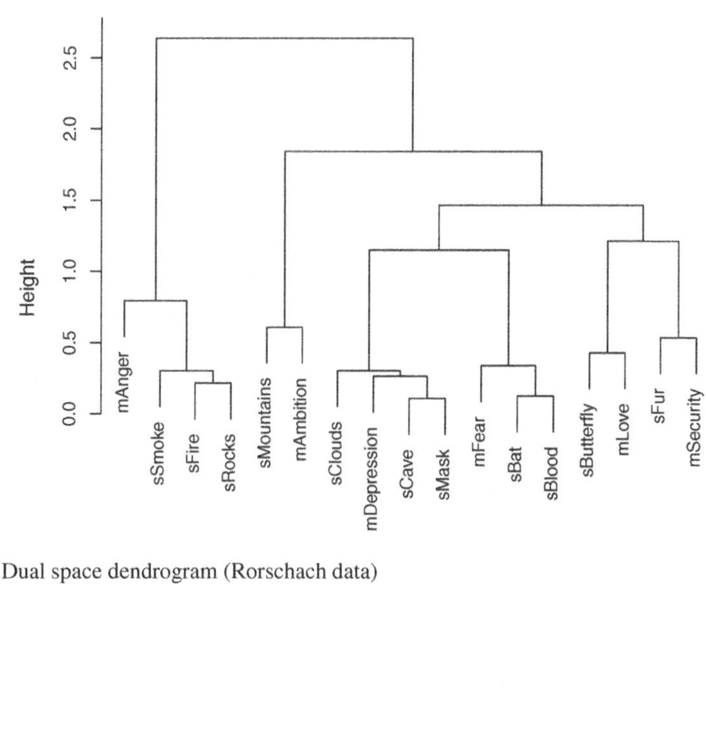

Fig. 6.1 Dual space dendrogram (Rorschach data)

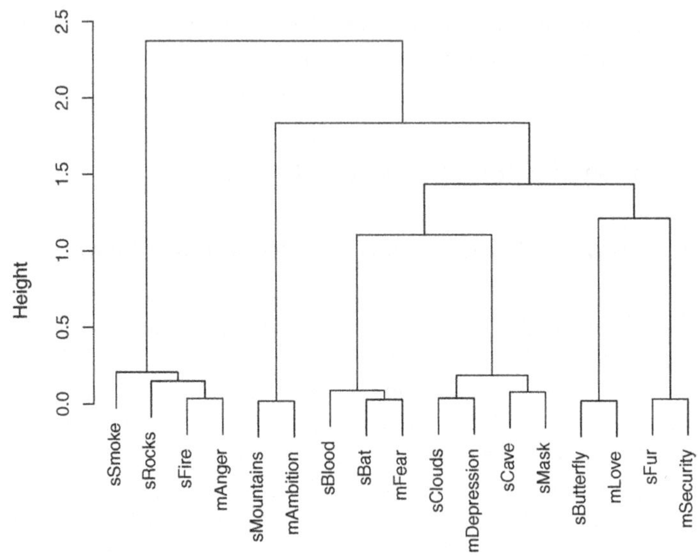

Fig. 6.2 Contingency space dendrogram (Rorschach data)

6.4.2 Hierarchical Cluster Analysis: Barley Data

The super-distance matrices of barley varieties and planting locations are shown in Table 6.10 for dual space, and Table 6.11 for contingency space. Since the number of rows is equal to the number of columns, total space is equal to dual space. Suplementary space distances are presented on Table 6.12.

The dendrograms are shown in Figs. 6.3 and 6.4. Noting that there is only a 22.8% discrepancy between dual space and contingency space, we expect that the clustering results from dual space and contingency space may not be substantial. Let us compare the two dendrograms. We can identify the following identical clusters for dual space and contingency space: (*Hanchen*; *St.Paul*), (*Manchria*; *Ithaca*), and (*Coast*; *Davis*, *Arlington*). We also see *Meloy* is quite isolated from the others in dual space and contingency space. The other clusters depend on space, which is due to the contributions of supplementary space, that is, whether this space is included or excluded.

6.4.3 Partitioning Cluster Analysis: Rorschach Data

The method creates an initial partitioning of variables and then uses an iterative relocation algorithm to relocate variables from one cluster to another until the optimization criterion is satisfied. We will use the `pam` algorithm described in Kaufman

Table 6.10 Dual Space (Barley data)

	Co	Ha	Wh	Ma	Ga	Me	Arl	Ith	StP	Moc	Mor	Dav
Co	0	0.9489	0.6048	1.0105	3.1197	8.0978	0.3466	0.8146	0.8269	0.7945	0.8066	0.4341
Ha	0.9489	0	1.027	1.4326	3.5418	8.5198	1.2548	1.1992	0.3335	1.2626	1.2241	1.1509
Wh	0.6048	1.027	0	1.0886	3.1977	8.1758	0.9166	0.9604	1.0052	0.5964	0.2217	0.8114
Ma	1.0105	1.4326	1.0886	0	3.6034	8.5815	1.3249	0.2091	1.4102	1.3118	1.2966	1.3079
Ga	3.1197	3.5418	3.1977	3.6034	0	10.691	3.2799	3.3592	3.3172	2.7235	3.4058	3.4053
Me	8.0978	8.5198	8.1758	8.5815	10.691	0	8.2826	8.4534	8.5048	8.3263	8.3838	7.53
Arl	0.3466	1.2548	0.9166	1.3249	3.2799	8.2826	0	1.1999	1.2514	1.2186	1.1304	1.1417
Ith	0.8146	1.1992	0.9604	0.2091	3.3592	8.4534	1.1999	0	1.2893	1.2567	1.1685	1.1797
StP	0.8269	0.3335	1.0052	1.4102	3.3172	8.5048	1.2514	1.2893	0	1.3081	1.2199	1.2311
Moc	0.7945	1.2626	0.5964	1.3118	2.7235	8.3263	1.2186	1.2567	1.3081	0	1.1872	1.1984
Mor	0.8066	1.2241	0.2217	1.2966	3.4058	8.3838	1.1304	1.1685	1.2199	1.1872	0	1.1102
Dav	0.4341	1.1509	0.8114	1.3079	3.4053	7.53	1.1417	1.1797	1.2311	1.1984	1.1102	0

Table 6.11 Contingency space (Barley data)

	Co	Ha	Wh	Ma	Ga	Me	Arl	Ith	StP	Moc	Mor	Dav
Co	0	0.8011	0.5402	0.9103	1.9892	4.617	0.0976	0.7607	0.7521	0.6239	0.7124	0.2585
Ha	0.8011	0	0.913	1.2615	2.325	5.1509	1.0996	1.0929	0.0065	1.0436	1.0537	0.9268
Wh	0.5402	0.913	0	1.0062	2.0344	4.8408	0.7827	0.8997	0.9141	0.4128	0.0547	0.6439
Ma	0.9103	1.2615	1.0062	0	2.4427	5.2646	1.1636	0.0082	1.3115	1.0971	1.1586	1.1041
Ga	1.9892	2.325	2.0344	2.4427	0	6.3738	2.1292	2.3411	2.2313	0.6314	2.7148	2.3991
Me	4.617	5.1509	4.8408	5.2646	6.3738	0	5.9734	5.1604	5.3306	4.6908	5.3181	2.7428
Arl	0.0976	1.0996	0.7827	1.1636	2.1292	5.9734	0	1.0023	1.0087	0.8479	0.9338	0.6612
Ith	0.7607	1.0929	0.8997	0.0082	2.3411	5.1604	1.0023	0	1.1349	0.9937	1.0514	0.9627
StP	0.7521	0.0065	0.9141	1.3115	2.2313	5.3306	1.0087	1.1349	0	1.0068	1.0639	0.9388
Moc	0.6239	1.0436	0.4128	1.0971	0.6314	4.6908	0.8479	0.9937	1.0068	0	0.7598	0.8068
Mor	0.7124	1.0537	0.0547	1.1586	2.7148	5.3181	0.9338	1.0514	1.0639	0.7598	0	0.8397
Dav	0.2585	0.9268	0.6439	1.1041	2.3991	2.7428	0.6612	0.9627	0.9388	0.8068	0.8397	0

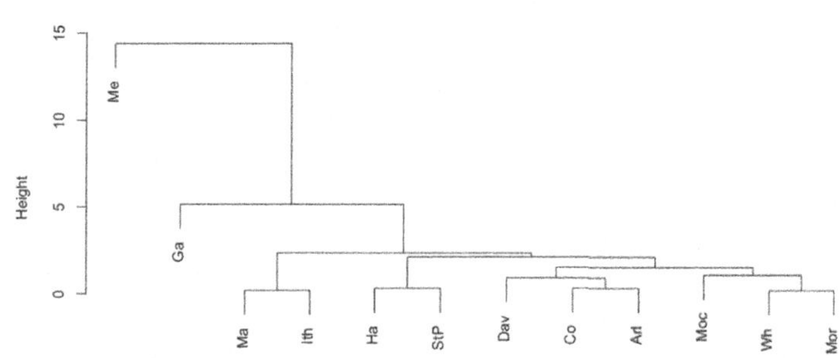

Fig. 6.3 Dual space dendrogram (Barley data)

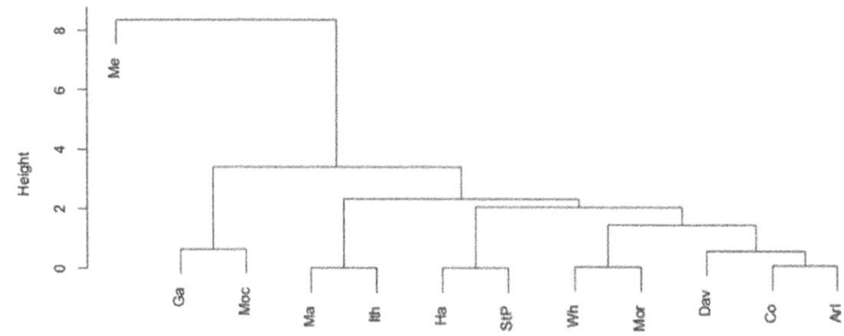

Fig. 6.4 Contingency space dendrogram (Barley data)

Table 6.12 Supplementary space (Barley data)

	Co	Ha	Wh	Ma	Ga	Me	Arl	Ith	StP	Moc	Mor	Dav
Co	0	0.1479	0.0646	0.1002	1.1305	3.4808	0.2490	0.0539	0.0748	0.1706	0.0942	0.1756
Ha	0.1479	0	0.114	0.1711	1.2167	3.369	0.1553	0.1063	0.3270	0.2190	0.1704	0.2241
Wh	0.0646	0.114	0	0.0824	1.1633	3.3351	0.1339	0.0607	0.0910	0.1836	0.1670	0.1675
Ma	0.1002	0.1711	0.0824	0	1.1607	3.3169	0.1613	0.2009	0.0988	0.2148	0.1381	0.2037
Ga	1.1305	1.2167	1.1633	1.1607	0	4.3168	1.1507	1.0181	1.0859	2.0921	0.6909	1.0063
Me	3.4808	3.369	3.3351	3.3169	4.3168	0	2.3092	3.2930	3.1742	3.6355	3.0657	4.7872
Arl	0.249	0.1553	0.1339	0.1613	1.1507	2.3092	0	0.1976	0.2426	0.3707	0.1966	0.4805
Ith	0.0539	0.1063	0.0607	0.2009	1.0181	3.293	0.1976	0	0.1544	0.263	0.1171	0.217
StP	0.0748	0.327	0.091	0.0988	1.0859	3.1742	0.2426	0.1544	0	0.3013	0.1561	0.2923
Moc	0.1706	0.219	0.1836	0.2148	2.0921	3.6355	0.3707	0.263	0.3013	0	0.4274	0.3916
Mor	0.0942	0.1704	0.167	0.1381	0.6909	3.0657	0.1966	0.1171	0.1561	0.4274	0	0.2705
Dav	0.1756	0.2241	0.1675	0.2037	1.0063	4.7872	0.4805	0.217	0.2923	0.3916	0.2705	0

and Rousseeuw (1990) because it is known to be a more robust version of the popular K-means clustering method. As is known, one problem for the users of this clustering procedure is that we must specify the number of clusters. Since there is no optimal method to determine the number of clusters, we will show a few examples for this demonstration. In both data sets, the smaller number of rows or columns is 6, which seems to suggest a reasonable number of clusters is 6. However, we will also examine 4 and 5 clusters.

Table 6.13 shows results when we specify the number of clusters to be 4, 5, and 6. The most dramatic finding here is the fact that cluster formations are quite different whether we use dual space or contingency space. Out of 4, 5, and 6 clusters, it looks as though 6 clusters are the most reasonable—recall that in hierarchical clustering one example of the six cluster solution provided identical results for both dual space and contingency space. The major difference between hierarchical clustering and partitioning is the fact that even for the case of 6 clusters, there are differences in partitioning between dual space and contingency space. We wonder why: Is this due to the point that partitioning is more precise in reflecting the contribution of supplementary space to clustering? Furthermore, partitioning results do not appear

Table 6.13 Partitioning in dual space and contingency space (Rorschach data). DUAL: dual space; CONT: contingency space

r/c	Name	K = 4		K = 5		K = 6	
		DUAL	CONT	DUAL	CONT	DUAL	CONT
Mood	Ambition	3	4	4	5	6	6
Mood	Anger	2	3	2	4	4	4
Mood	Depression	1	1	1	3	2	3
Mood	Fear	1	1	1	1	1	1
Mood	Love	1	2	3	2	4	2
Mood	Security	1	2	1	2	4	5
Symbol	Bat	1	1	1	1	1	1
Symbol	Blood	2	1	2	1	1	1
Symbol	Butterfly	1	2	3	2	2	2
Symbol	Cave	1	1	1	3	2	3
Symbol	Clouds	1	1	1	3	2	3
Symbol	Fire	2	3	2	4	1	4
Symbol	Fur	3	2	3	2	3	5
Symbol	Mask	1	1	1	3	4	3
Symbol	Mountains	3	4	4	5	5	6
Symbol	Rocks	2	3	2	4	2	4
Symbol	Smoke	4	3	5	4	6	4

as clear as the results of hierarchical clustering: dendrograms are more informative since they also indicate distances between clusters.

6.4.4 *Partitioning Cluster Analysis: Barley Data*

Since the number of rows and that of columns is 6, we will again consider $k = 4$, $k = 5$, and $k = 6$ clusters in dual space and contingency space. The results are summarized in Table 6.14.

For this data set, the correct number of clusters seems to be six, for the results are basically the same as those from hierarchical clustering. Considering rather confusing results from the cases of 4 and 5 clusters, one conclusion we can make is that it is very important to identify a reasonable number of clusters in partitioning, which may not be an easy task.

Table 6.14 Partitioning into 4, 5, and 6 clusters (Barley data). DUAL: dual space; CONT: contingency space

r/c	Names	K = 4		K = 5		K = 6	
		DUAL	CONT	DUAL	CONT	DUAL	CONT
Location	Arl	1	1	1	1	1	1
Location	Dav	1	1	1	1	1	1
Location	Ith	2	3	3	4	4	4
Location	Moc	1	2	1	3	3	3
Location	Mor	1	1	1	1	3	3
Location	StP	1	1	2	2	2	2
Variety	Co	1	1	1	1	1	1
Variety	Ga	3	2	4	3	5	5
Variety	Ha	1	1	2	2	2	2
Variety	Ma	2	3	3	4	4	4
Variety	Me	4	4	5	5	6	6
Variety	Wh	1	2	1	3	3	3

6.5 Cluster Analysis of Between-Set Relations

As Nishisato (2014) indicated, it would be more direct to cluster the between-set distance matrix since our main interest lies in the analysis of row-column relations of the contingency table. There are two major obstacles for clustering of the between-set matrix: (1) the input matrix may not be square, and (2) most importantly, even if the matrix is square, it is not *symmetric*. Thus, this is where our framework of modified data matrices (CMDM) can play an important role in using hierarchical and partitioning algorithms: by selecting the between-set distance sub-matrix of the super-distance matrix for clustering, we can use both hierarchical clustering algorithm and partitioning algorithm.

6.5.1 Hierarchical Cluster Analysis of Rorschach Data (UTC)

The modification of the input matrix is done this time with $a = c = d = 100$ and $b = 1$. Once we have transformed the super-matrix of distances, we apply both hierarchical clustering and partitioning methods to the modified super-distance matrix.

Let us look at the two dendrograms. In both dual space (Fig. 6.5) and contingency space (Fig. 6.1), we see the following matching clusters: ($Love$: $Butterfly$), ($Security$: Fur), ($Fear$: Bat, $Blood$), ($Ambition$: $Mountains$), and ($Anger$: $Rocks$, $Fire$). We then see differences with remaining moods and inkblots, a reflec-

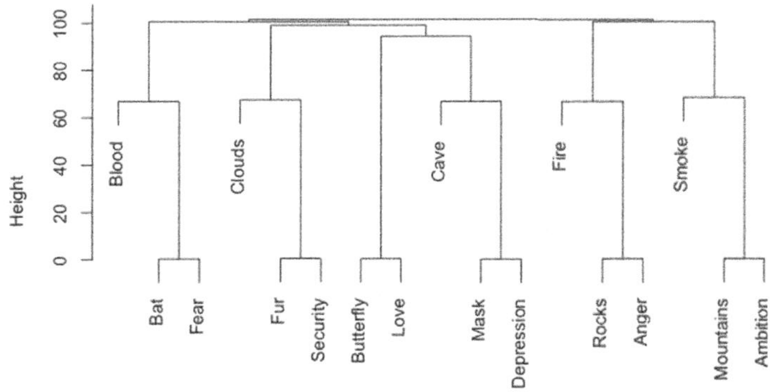

Fig. 6.5 Dendrogram of dual space (UTC: Rorschach data)

tion of the contributions of supplementary space. Again, hierarchical clustering offers relatively easy and reasonable results.

6.5.2 *Hierarchical Cluster Analysis of Barley Data (UTC)*

We set the selection constants as $a = c = d = 100$ and $b = 1$. The clustering results of hierarchical clustering are summarized in the dendrogram.

The dendrograms for dual space and contingency space are as shown in Figs. 6.6 and 6.7. The dendrograms are identical. Looking at the cluster formation, can we conclude that our UTC works well?

6.5.3 *Partitioning Cluster Analysis: Rorschach Data and Barley Data (UTC)*

We again consider 4, 5, and 6 clusters. Partitioning results of these two data sets are summarized in Tables 6.15 and 6.16.

Partitioning of the modified matrix into 6 clusters yields identical results to the one from the super-distance matrix, perhaps an indication of preference for the UTC over dealing with the larger square matrix. However, when we look at other numbers of clusters, these two sets of results are quite different, suggesting again that the partitioning method of cluster analysis may not be a good alternative to joint graphical display—what if the investigator decides to use a wrong number of clusters?

The results from partitioning are different from the results of hierarchical clustering. Tentatively speaking, the partitioning method may not be a good alternative to joint graphical display.

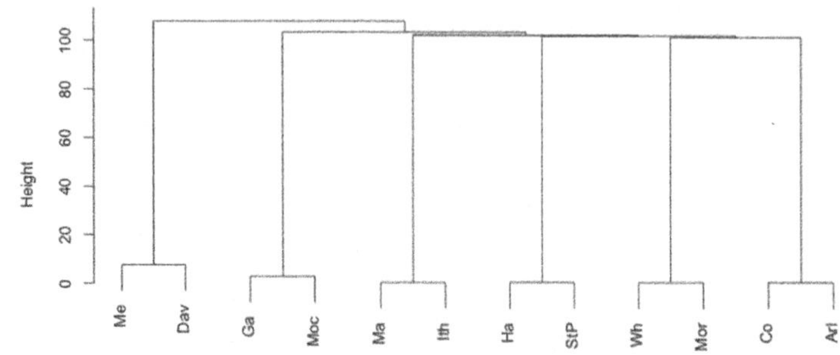

Fig. 6.6 Dual space dendrogram (UTC: Barley data)

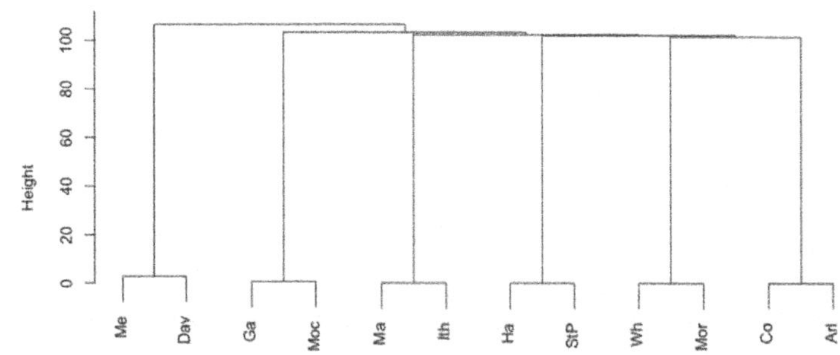

Fig. 6.7 Contingency space dendrogram (UTC: Barley data)

6.5.4 Effects of Constant Q for UTC on Cluster Formation

In the above examples, we used Q = 100, but what if we use a different number, while maintaining the number definitely larger than any of the elements in the sub-matrices to be ignored? We have numerically verified that UTC clusters remain identical so long as the smallest elements in the ignored sub-matrices are larger than the largest element in the target matrix for clustering.

6.6 Overlapping Versus Non-overlapping Clusters

None of the methods we used can handle overlapping clusters. The question is how to examine the data to see if overlapping clusters are more appropriate than non-overlapping clusters. This is a problem of fundamental importance, yet the methods we used here do not allow overlapping clusters.

From the user's point of view, this is a legitimate question because we are interested in finding data structure, not a fictionally imposed structure, peculiar to the chosen method of analysis. We know the method of bi-clustering, which we did not use for the reasons stated earlier, can accommodate overlapping clusters. Another direct

Table 6.15 Partitioning cluster analysis (UTC: Rorschach data)

r/c	Name	K = 4		K = 5		K = 6	
		DUAL	CONT	DUAL	CONT	DUAL	CONT
Mood	Ambition	2	4	3	5	4	6
Mood	Anger	2	2	3	4	4	4
Mood	Depression	1	1	1	3	2	3
Mood	Fear	2	3	3	1	1	1
Mood	Love	2	3	3		4	2
Mood	Security	2	3	3	2	4	5
Symbol	Bat	1	1	1	1	1	1
Symbol	Blood	1	2	1	1	1	1
Symbol	Butterfly	1	3	1	2	2	2
Symbol	Cave	1	1	1	3	2	3
Symbol	Clouds	1	1	1	3	2	3
Symbol	Fire	1	2	1	4	1	4
Symbol	Fur	1	1	2	3	3	5
Symbol	Mask	2	1	3	3	4	3
Symbol	Mountains	3	4	4	5	5	6
Symbol	Rocks	1	2	1	4	2	4
Symbol	Smoke	4	2	5	4	6	4

method which can respond to this question is Nishisato's filtering method (2014), where relatively large distances are gradually dropped from the clustering data step by

Table 6.16 Partitioning cluster analysis (UTC: Barley data)

r/c	Name	K = 4		K = 5		K = 6	
		DUAL	CONT	DUAL	CONT	DUAL	CONT
Location	Arl	1	1	1	1	1	1
Location	Dav	1	1	1	1	1	1
Location	Ith	2	3	2	4	4	4
Location	Moc	1	2	3	3	3	3
Location	Mor	1	1	3	1	3	3
Location	StP	1	1	1	2	2	2
Variety	Co	1	1	1	1	1	1
Variety	Ga	3	2	4	3	5	5
Variety	Ha	2	2	2	2	2	2
Variety	Ma	2	3	2	4	4	4
Variety	Me	4	4	5	5	6	6
Variety	Wh	2	2	3	3	3	3

step, revealing tighter and tighter clustering cores. In applying his method, Nishisato (2012) and Clavel and Nishisato (2020) have shown that overlapping clusters are more appropriate for our barley data.

This is a lesson to learn that the beauty of analytical results alone does not verify the appropriateness of an analytical tool. It seems that any transformation of data, either analytical or graphical, should reflect the structure of the original data. We hope that both hierarchical clustering and partitioning method will further be developed so as to accommodate overlapping clusters.

6.7 Discussion and Conclusion

We have found a number of important points about this possible alternative for joint graphical display. Among others, the most important point is that cluster analysis does not have any problems over dimensionality of data and offers an excellent way of summarizing data in an interpretable way. The unique aspect of the current chapter is that we looked at cluster compositions of data in dual space and contingency space. Although dual space contains the entire information of row-column relations of the contingency table, the present demonstrations show that data analysis in contingency space yields a good depiction of data structure, indicating that clustering, particularly hierarchical cluster analysis, can be a practical way of describing data. Our main conclusions can be summarized as follows:

- Hierarchical clustering appears to provide a stable and interpretable picture of row-column relations in the contingency table than the partitioning method.
- Our newly proposed UCT provides a transformed super-distance matrix, which allows that any currently available clustering methods can handle clustering of rectangular, square, non-symmetric, or symmetric sub-matrix of the super-distance matrix.

6.8 Final Comments on Part 1

We have discussed topics related to joint graphical display. We now would like to look at some thoughts on the future of the current endeavour.

First, we showed our successful derivation of exact Euclidean coordinates for both rows and columns of the contingency table in common space. In this process, we presented our theory of quantification space, where total space is partitioned into dual space (contingency space + supplementary space) and residual space. This theory clarified the domain of meaningful quantification space. The perennial problem of joint graphical display is finally solved through quantification of dual space, whereby the name "dual scaling" was fully justified, posing a question, at the same time, for possible justification of the name correspondence analysis. While the French plot was shown to provide an excellent approximation to the correct graphical representation of data, we were made keenly aware of the serious limitations of the graphical approach to the multidimensional configuration of data.

As an alternative to joint graphical display, we discussed cluster analysis, and demonstrated that cluster analysis is a very promising alternative. We developed the framework which makes it possible to subject any rectangular or square non-symmetric distance sub-matrix of the super-distance matrix to the traditional clustering algorithm (most of these were developed only for square symmetric data matrices). Let us briefly summarize our findings with comments for future directions of research.

- Quantification of the contingency table should ideally be carried out in dual space.
- French plot in contingency space offers a good practical approximation to the plot in dual space.
- Joint graphical display in dual space leaves a great deal of work to be done, namely the main task is how to display relations in more than three dimensions.
- In contrast, hierarchical clustering provides an excellent means of summarizing multidimensional configurations into clusters.
- Cluster analysis has also hinted that contingency space is a convenient and good approximation to the structure of dual space if one wants to simplify the analysis.
- One should try to minimize residual space as a research strategy, that is, try to consider the equal numbers for rows and columns so long as it is possible.
- Our newly proposed UCT provides a transformed super-distance matrix, which allows any currently available clustering methods that can handle clustering of rectangular, square, non-symmetric, or symmetric sub-matrix of the super-distance matrices.
- We need to develop data analytical tools which reflect our data structures. In particular, we must develop a hierarchical clustering method which can handle overlapping clusters.

In Part 1, we presented joint graphical display and cluster analysis as alternatives for summarizing quantification results. Although our discussions are limited to the contingency table and its response-pattern format, there are many other kinds of data, categorical and continuous, that require similar consideration of information retrieval and practical ways of representing the analytical results. It is hoped that Part 1 serves as a starting point for a further endeavour for developing useful and valid methods of data analysis and useful ways of representing the analytical results, be it graphical display or cluster analysis. Please also keep in mind the importance of the theory of quantification space, where dual space should be the central focal point of our interest in row-column relations of the contingency table.

References

Clavel, J. & Nishisato, S. (2008). Joint analysis of within-set and between-set distances. In Shigemasu, K. e. a. (Eds.), *New trends in psychometrics* (pp. 41–50). Tokyo: Universal Academy Press.

Clavel, J. G. & Nishisato, S. (2012). *Challenges at the interface of data analysis, computer science, and optimization*, chapter Reduced versus complete space configurations in total information analysis (pp. 91–99). Berlin: Springer.

Clavel, J. G. & Nishisato, S. (2020). From joint graphical display to bi-modal clustering: [2] dual space versus total space. In Imaizumi, T., Okada, A., Miyamoto, S., Sakaori, F., Yamamoto, Y., & Vichi, M. (Eds.), *Advanced Studies in Classification and Data Science* (pp. 131–143). Singapore: Springer Singapore.

Garmize, L., & Rycklak, J. (1964). Role-play validation of socio-cultural theory of symbolism. *Journal of Consulting Psychology, 28,* 107–115.

Kaufman, L., & Rousseeuw, P. (1990). *Finding Groups in Data*. New York: Wiley.

Nishisato, S. (1984). Forced classification: A simple application of a quantification method. *Psychometrika, 49*(1), 25–36.

Nishisato, S. (1994). *Elements of dual scaling*. Hillsdale, NY: Lawrence Erlbaum Associates, Inc., Publishers.

Nishisato S. (2012) Quantification Theory: Reminiscence and a Step Forward. In Gaul W., Geyer-Schulz A., Schmidt-Thieme L., & Kunze J. (Eds.), *Challenges at the interface of data analysis, computer science, and optimization*. Studies in Classification, Data Analysis, and Knowledge Organization. Berlin, Heidelberg: Springer.

Nishisato, S. (2014). Structural representation of categorical data and cluster analysis through filters. In Gaul, W., Geyer-Schulz, A., Baba, Y., & Okada, A. (Eds.), *German-Japanese interchange of data analysis results* (pp. 81–90). Cham: Springer International Publishing.

Nishisato, S., & Baba, Y. (1999). On contingency, projection and forced classification of dual scaling. *Behaviormetrika, 26*(2), 207–219.

Nishisato, S., & Clavel, J. (2008). Interpreting data in reduced space: A case of what is not what in multidimensional data analysis. In Shigemasu, K. e. a. (Eds.), *New trends in psychometrics* (pp. 357–366). Tokyo: Universal Academy Press.

Nishisato, S., & Lawrence, D. (1989). *Multiway data analysis*, chapter Dual Scaling of multiway data matrices: several variants (pp. 317–326). Amsterdam: Elsevier Science Publisher.

Stebbins, G. L. (1950). *Variation and evolution in plants*. New York: Columbia University Press.

Ward, J. H. (1963). Hierarchical grouping to optimize an objective function. *Journal of the American Statistical Association, 58*(301), 236–244.

Part II
Scoring Strategies and the Graphical Display

Preface

While the preface to Part I was written from the perspective of Shizuhiko Nishisato, we have taken a slightly different tact for the preface of Part II. We have presented a separate preliminary discussion of our experiences and how they impacted the following chapters.

Notes on Eric J. Beh

I was never formally taught any form of categorical data analysis. I didn't have the opportunity to take any courses on it while I was studying. Instead, I learnt it by writing a lot (and refining my style) and reading even more. My first experience of learning anything about categorical data analysis was in 1994 when I was in my Honours year and I was asked to write a thesis on a 1984 book that was in the University of Wollongong (Australia) library called *Theory and Applications of Correspondence Analysis* by Michael Greenacre. At the time I had no idea what correspondence analysis was, nor did I know of the importance of Michael Greenacre's impact. Unlike my fellow authors—José Clavel and Rosaria Lombardo—I didn't have the luxury of having a scaling or correspondence analysis "master" as a supervisor (Pam Davy) but she was wonderful and instilled in me the importance of curiosity and always kept asking me questions. I also had the benefit of John Rayner as my co-supervisor and, along with Pam, he expected that I constantly ask questions of myself. During my Honours year, I was exposed (briefly) to the work of Ludovic Lebart, Jan de Leeuw, and Shizuhiko Nishisato. I had also been introduced to some of the work of Alan Agresti, Leo Goodman, (Henry) Oliver Lancaster (who was John's supervisor at the University of Sydney), and Shelby Haberman—universal masters in categorical data analysis. Their work felt more "natural" to me as a mathematical statistician than the dependence on geometry, extensive use of matrices, and dimension reduction that scaling/correspondence analysis demanded of me. Despite this, I felt increasingly curious and the fact that there was pretty much no one in Australia that I could find (or that I was exposed to) that knew a great deal about correspondence analysis meant that later in my studies I would be looking predominantly abroad, especially to Europe. Part of the focus of my honours thesis was to investigate how ordered categories were treated (a topic that was motivated by a single question at a bad presentation I did during my Honours year). After Honours, I liked what I had learned about correspondence analysis and the seed to my first paper (in 1997) had started to germinate so I commenced my Ph.D. studies in the middle of 1995. I picked up where I left Honours and I had learned of a range of techniques which all appeared to enforce an order of the scales along a single dimension. Then I came across a 1975 paper by Shizuhiko Nishisato and P. S. Arri titled *Nonlinear programming approach to optimal scaling of partially ordered categories* that was published in *Psychometrika.* While confining their attention to the scoring of a particular case of ordered categories, it was there that I learned a lot more about the importance of scaling and of reciprocal averaging. I immersed myself in Hill's 1974 landmark paper, and better understood the link between scaling, the eigendecomposition of a matrix (which, at the time, I barely remembered from my undergraduate days doing linear algebra), and singular value decomposition.

While I may be reminiscing more than providing an overview of the scope, direction, and aims of this part of the book, it does have some relevance. Firstly, my origins and therefore perceptions of scaling and correspondence analysis may not necessarily blend naturally with those who have been more "formally" trained. So my philosophies tend to be more statistical than most and so my view of correspondence analysis is less geometric; of course, there remains a strong geometric element. As such, my involvement in Part II of this book has been to not just continue the wonderful journey I've had over the past decade or so in working with Rosaria, but to provide a bit of a glimpse into the inner workings of reciprocal averaging and its role in correspondence analysis. By virtue of my influencers, it strongly and unashamedly draws parallels with that of Hill (1974). However, I have also tried to think a little bit more "outside the box"—and this is what motivates Chap. 8 (for good or bad). Of course, the idea of scaling leads nicely to the visual depiction of the association between categorical variables of which correspondence analysis plays a fundamentally important role and where Rosaria's and my interests greatly intersect. She revealed to me years ago the unspoken nature of classical correspondence analysis which deals exclusively with symmetrically associated variables while the Italians (through Luigi D'Ambra and Carlo Lauro) had developed a variant called non-symmetrical correspondence analysis to accommodate for the asymmetric association. The work I had done dealing with ordered categorical variables using orthogonal polynomials blended nicely with these approaches and Chap. 10 provides an overview of some of them; although more details on these can be found in Beh and Lombardo (2014; 2020b). There are some other aspects that we also investigate and Rosaria will now provide her discussion and visions of the chapters in Part II.

Eric J. Beh,
University of Newcastle, Australia.

Notes on Rosaria Lombardo

My background is fairly similar to Eric's. My first experience of learning about correspondence analysis was in 1990 when I was in my final year of a five-year degree and I was asked to write a thesis on a variant of simple correspondence analysis called *Non-Symmetrical Correspondence Analysis* (NSCA). At that time, I had the opportunity to interact with two correspondence analysis "masters"—Luigi D'Ambra and Natale Carlo Lauro. They motivated me to commence my Ph.D. on issues concerned with computational statistics and its applications at the Department of Mathematics and Statistics of the University Federico II in Naples (Italy). So, I continued my studies and focussed on NSCA, dependence models, quantification theory, and graphical displays. At that time, Carlo Lauro invited Jan van Rijckevorsel to the Department of Mathematics and Statistics (Naples) to give a series of lectures on reciprocal averaging and nonlinear principal component analysis (for numerical and categorical variables) which instilled in me the idea of alternative perspectives of correspondence analysis; such perspectives were certainly different from the French approach to data analysis that I had been exposed to.

During my Ph.D., I visited Pieter Kroonenberg (University of Leiden, The Netherlands) and André Carlier (University of Toulouse "Paul Sabatier", France) with the aim of learning more about graphical displays, correspondence analysis, and three-way data analysis. They were both wonderful and stimulated me to ask a range of theoretical and practical questions. During that time, Pieter and André were working together on biplots for three-way correspondence analysis and they both "opened my mind" to different geometric "maps" when portraying the symmetrical and non-symmetrical association of categorical data. Like Eric, I was exposed to the work of Jean-Paul Benzécri, Michael Greenacre, (Henry) Oliver Lancaster, Ludovic Lebart, and Shizuhiko Nishisato, but I was also greatly influenced by those from the Dutch school of data analysis led by Jan de Leeuw, including Peter van der Heijden, Willem J. Heiser, Jaqueline J. Meulman, Jan van Rijckevorsel, and (of course) Pieter. Their work fascinated, intrigued, and motivated me to learn more about data coding and quantification theory for categorical variables.

After my Ph.D. course, I had the fortune to get to know another "beautiful mind", Jean-François Durand (University of Montpellier, France), who helped me enormously to deepen my knowledge on nonlinear polynomials for data coding and partial least squares. Furthermore, over the last few years, my interest in issues concerned with three-way data analysis led me to broaden some of the work done by Yoshio Takane. While I only knew Yoshio virtually (through many email conversations), I knew him to be charming and brilliant and he enlightened me about the different ways of partitioning symmetric and asymmetric three-way indexes as discussed in Lombardo et al. (2020b).

Therefore, my perceptions of scaling and correspondence analysis tend to be both geometric and statistical and thanks to the wonderful journey that I've had over the last 15 years (or so) with Eric; many other mathematical facets of coding and issues concerned with the quantification of categorical data have enlightened me. Eric's and my interests in scaling and quantification theory have motivated our discussion of reciprocal averaging in Chap. 7 and to propose some variants to it in Chap. 8. My experiences in visualization techniques inspired the discussion of biplots for numerical data in Chap. 9. Eric's extensive work with ordered categorical variables and my long exposure to non-symmetrical correspondence analysis and multi-way data quantification have been united in Chap. 10 where we discuss the role of biplots for ordered and nominal two-way and multi-way contingency tables. Eric and I would also like to thank Pieter for his role in the development of these ideas which are more fully explored in Lombardo et al. (2016, 2020a). Furthermore, some new work that Eric and I have undertaken explores the implications of dealing with over-dispersed categorical data and this has motivated our discussion in Chap. 11. This chapter also expands upon what we have recently published Beh and Lombardo (2020a).

Rosaria Lombardo,
University of Campania L. Vanvitelli, Italy.

References

Beh, E. J., & Lombardo, R. (2014). *Correspondence analysis: Theory, practice and new strategies*. Chichester: Wiley.

Beh, E. J., & Lombardo, R. (2020a). Five strategies for accomodating overdispersion in correspondence analysis. In Imaizumi, T., Okada, A., Miyamoto, S., Sakaori, F., Yamamoto, Y., & Vichi, M. (Eds.), *Advanced studies in classification and data science* (pp. 117–129). Singapore: Springer.

Beh, E. J., & Lombardo, R. (2021). *An introduction to correspondence analysis*. Chichester: Wiley.

Hill, M. (1974). Correspondence analysis: a neglected multivariate technique. *Journal of the Royal Statistical Society (Series, C)*, 23: 340—354.

Lombardo, R., Beh, E. J., & Kroonenberg, P. M. (2016). Modelling trends in ordered correspondence analysis using orthogonal polynomials. *Psychometrika*, 81:325—349.

Lombardo, R., Beh, E. J., & Kroonenberg, P. M. (2020a). Symmetrical and non-symmetrical variants of three-way correspondence analysis for ordered variables. *Statistical Science*, page (in press).

Lombardo, R., Takane, Y., & Beh, E. J. (2020b). Familywise decompositions of Pearson's chi-squared statistic in the analysis of contingency tables. *Advances in Data Analysis and Classification*, 14: 629—249.

Chapter 7
Scoring and Profiles

7.1 Introduction

At the heart of visually summarising the association between two categorical variables using techniques such as correspondence analysis is the *profile*. Simply put, a profile is merely the relative distribution of cell frequencies for a row or column. In this chapter, we shall be defining a row profile and a column profile and show how one can determine scores for each category so that some understanding of the differences (or similarities) between rows (and columns) can be made in terms of these profiles. The aim of doing so is that two row scores, say, that are similar to each other reflect the fact their row categories have similar profiles, while different scores highlight those row profiles that are different. The correlation between the row and column scores can also be determined and helps to quantify the association that exists between the categorical variables. By calculating the scores in the manner that we describe ensures that the correlation between them is maximized which leads to the best possible visual representation of the row and column categories in a low-dimensional space. The low-dimensional space that we shall describe is the *biplot* and will be the focus of our discussion in Chaps. 9 and 10.

There are numerous ways in which the row and column scores can be determined and so various strategies have been developed over the years that lead to the same result. For example, such strategies include, but are not limited to reciprocal averaging, canonical correlation analysis, dual scaling, and homogeneity analysis. Rather than giving an exhaustive overview of the various techniques, we shall instead confine our attention to the methods of reciprocal averaging and canonical correlation analysis. Therefore, we must stress that this chapter does not present anything that has not appeared elsewhere in the statistical (or allied) literature. Rather, our description of reciprocal averaging is aligned with that described by Hill (1974) while our discussion of canonical correlation analysis is consistent with that given by Hirschfeld (1935). However, the description we make of these techniques in this chapter is important for our discussion of variants of reciprocal averaging that we make in Chap. 8.

S. Nishisato et al., *Modern Quantification Theory*,
Behaviormetrics: Quantitative Approaches to Human Behavior 8,
https://doi.org/10.1007/978-981-16-2470-4_7

7.2 Profiles

For a two-way contingency table of size $m \times n$, we define the ith row profile to be the distribution of the relative cell frequencies such that

$$\left(\frac{f_{i1}}{f_{i\bullet}}, \frac{f_{i2}}{f_{i\bullet}}, \ldots, \frac{f_{ij}}{f_{i\bullet}}, \ldots, \frac{f_{in}}{f_{i\bullet}}\right).$$

We divide each cell frequency by the row total since not all rows will have the same marginal frequency. This profile may be expressed in terms of joint and marginal proportions by dividing the joint and marginal frequencies by the sample size f_t yielding

$$\left(\frac{p_{i1}}{p_{i\bullet}}, \frac{p_{i2}}{p_{i\bullet}}, \ldots, \frac{p_{ij}}{p_{i\bullet}}, \ldots, \frac{p_{in}}{p_{i\bullet}}\right). \tag{7.1}$$

The profile for the jth column is defined as the distribution of the relative cell frequencies of that column and is

$$\left(\frac{f_{1j}}{f_{\bullet j}}, \frac{f_{2j}}{f_{\bullet j}}, \ldots, \frac{f_{ij}}{f_{\bullet j}}, \ldots, \frac{f_{mj}}{f_{\bullet j}}\right) \tag{7.2}$$

which can be alternatively, but equivalently, expressed as

$$\left(\frac{p_{1j}}{p_{\bullet j}}, \frac{p_{2j}}{p_{\bullet j}}, \ldots, \frac{p_{ij}}{p_{\bullet j}}, \ldots, \frac{p_{mj}}{p_{\bullet j}}\right).$$

Centred versions of the ith row profile and jth column profile can also be considered. For such versions, the centring is undertaken with respect to the expected profile if the row and column variables are independent. Doing so ensures that when it comes to visualising the association between the row and column variables the origin of the resulting plot has a clear and meaningful interpretation. Since the expected value of f_{ij} under the independence is $f_{i\bullet}f_{\bullet j}/f_t$ then the ith centred row profile is

$$\left(\frac{f_{i1}}{f_{i\bullet}} - f_{\bullet 1}, \frac{f_{i2}}{f_{i\bullet}} - f_{\bullet 2}, \ldots, \frac{f_{ij}}{f_{i\bullet}} - f_{\bullet j}, \ldots, \frac{f_{in}}{f_{i\bullet}} - f_{\bullet n}\right)$$

or, in terms of the joint and marginal proportions, by

$$\left(\frac{p_{i1}}{p_{i\bullet}} - p_{\bullet 1}, \frac{p_{i2}}{p_{i\bullet}} - p_{\bullet 2}, \ldots, \frac{p_{ij}}{p_{i\bullet}} - p_{\bullet j}, \ldots, \frac{p_{in}}{p_{i\bullet}} - p_{\bullet n}\right).$$

By following a similar argument, the jth centred column profile can be expressed in terms of the joint and marginal frequencies such that

$$\left(\frac{f_{1j}}{f_{\bullet j}} - f_{1\bullet}, \frac{f_{2j}}{f_{\bullet j}} - f_{2\bullet}, \ldots, \frac{f_{ij}}{f_{\bullet j}} - f_{i\bullet}, \ldots, \frac{f_{mj}}{f_{\bullet j}} - f_{m\bullet}\right)$$

or, equivalently, in terms of their proportions by

$$\left(\frac{p_{1j}}{p_{\bullet j}} - p_{1\bullet}, \frac{p_{2j}}{p_{\bullet j}} - p_{2\bullet}, \ldots, \frac{p_{ij}}{p_{\bullet j}} - p_{i\bullet}, \ldots, \frac{p_{mj}}{p_{\bullet j}} - p_{m\bullet}\right).$$

7.3 The Method Reciprocal Averaging

7.3.1 An Overview

A feature of the row profiles, say, or their centred version, is that they may be used to help visually summarize those rows and columns that are similarly, or differently, distributed. For example, the ith centred row profile may be viewed as a point in n-dimensional space while the jth centred column profile may be viewed as a point in m-dimensional space - such spaces are referred to as a *cloud of points* in the correspondence analysis literature. When $m > 3$ and/or $n > 3$, viewing a cloud of points that consists of at least four dimensions is problematic. Therefore, we require a strategy that reduces the dimensionality of these two spaces so that one may easily view the comparison of the profiles and also reflect the association between the variables. One such strategy, and the approach we describe in this chapter, is the method of *reciprocal averaging*, which has also been referred to as the *method of reciprocal averages* throughout the psychology literature; we also provided an overview of this method in Sect. 3.2.1. Richardson and Kuder (1933) [p.36] first proposed a rating scale where

> ... each statement is scaled, i.e., it is given a quantitative value upon a scale with equal units in such a way that the endorsement or refusal to endorse a statement may be interpreted in quantitative terms

and based their method of scaling on Thurstone's (1928) *method of equal appearing intervals*. Horst (1935) then elaborated further on Richardson and Kuder's (1933) idea by demonstrating its applicability to a variety of different variables. Further developments continued to be made, especially in the psychology literature, for the following decades; one may refer to, for example, the contributions made by Mosier (1946), Mitzel (1954) and Lawshe and Harris (1958). However, the first rigorous mathematical treatment of reciprocal averaging was that of Hirschfeld (1935); see also Michailidis and de Leeuw (1984), Van Rijckevorsel (1987) [pp. 24–29] and Gifi (1990) [p. 107]. Perhaps the biggest impact that reciprocal averaging has had on the visualization of the association between categorical variables is that of Hill (1974). He demonstrated that the classic reciprocal averaging technique is in fact analogous to the scaling of variables required for the visualization of their association. With this in mind, we shall motivate our discussion of reciprocal averaging in this chapter on Hill's (1974) exposition. Interestingly, Welsh and Robinson (2005) [p. 144] credit Fisher and Mackenzie (1923) with the computational development of reciprocal averaging. There are indeed some parallels between the work of Hirschfeld (1935) and Fisher and Mackenzie (1923); both result in two "reciprocal averaging" equations that are similar in nature and the correlation of the resulting scores is also quite similar. However, Fisher and Mackenzie (1923) taking a least-squares approach to optimizing scores of a two-way ANOVA problem but do not impose any particular property, or constraint, on the resulting scores.

7.3.2 *Profiles*

The aim of reciprocal averaging, like that of dual scaling (and all of the analogous scaling techniques), is to determine the scores associated with the ith row and jth column categories so that we may view them in a low-dimensional space. Suppose the ith row score is denoted by x_{ik} and the jth column score is denoted by y_{ik}. These scores can be determined by finding the average row and column of their profiles such that

$$\rho_k x_{ik} = \left(\frac{p_{i1}}{p_{i\bullet}} - p_{\bullet 1}\right) y_{1k} + \cdots + \left(\frac{p_{in}}{p_{i\bullet}} - p_{\bullet n}\right) y_{nk} = \sum_{j=1}^{n} \left(\frac{p_{ij}}{p_{i\bullet}} - p_{\bullet j}\right) y_{jk} \tag{7.3}$$

and

$$\rho_k y_{jk} = \left(\frac{p_{1j}}{p_{\bullet j}} - p_{1\bullet}\right) x_{1k} + \cdots + \left(\frac{p_{mj}}{p_{\bullet j}} - p_{m\bullet}\right) x_{mk} = \sum_{i=1}^{m} \left(\frac{p_{ij}}{p_{\bullet j}} - p_{i\bullet}\right) x_{ik} \tag{7.4}$$

where x_{ik} and y_{jk} the following properties are most frequently imposed

$$\mathrm{E}\left(x_{ik}\right) = \sum_{i=1}^{m} p_{i\bullet} x_{ik} = 0 \qquad \mathrm{Var}\left(x_{ik}\right) = \sum_{i=1}^{m} p_{i\bullet} x_{ik}^2 = 1 \tag{7.5}$$

$$\mathrm{E}\left(y_{jk}\right) = \sum_{j=1}^{n} p_{\bullet j} y_{jk} = 0 \qquad \mathrm{Var}\left(y_{jk}\right) = \sum_{j=1}^{n} p_{\bullet j} y_{jk}^2 = 1\,. \tag{7.6}$$

These properties may be expressed in matrix form by

$$\mathrm{E}\left(\mathbf{x}_k\right) = \mathbf{r}^T \mathbf{x}_k = 0 \qquad \mathrm{Var}\left(\mathbf{x}_k\right) = \mathbf{x}_k^T \mathbf{D}_r \mathbf{x}_k = 1 \tag{7.7}$$

$$\mathrm{E}\left(\mathbf{y}_k\right) = \mathbf{c}^T \mathbf{y}_k = 0 \qquad \mathrm{Var}\left(\mathbf{x}_k\right) = \mathbf{y}_k^T \mathbf{D}_c \mathbf{y}_k = 1\,. \tag{7.8}$$

Note that, from Eqs. (7.5) and (7.6), (7.3) and (7.4) may be expressed as the two uncentred formulae of reciprocal averaging such that

$$\rho_k x_{ik} = \sum_{j=1}^{n} \left(\frac{p_{ij}}{p_{i\bullet}} \right) y_{jk}$$

and

$$\rho_k y_{jk} = \sum_{i=1}^{m} \left(\frac{p_{ij}}{p_{\bullet j}} \right) x_{ik}$$

or, in matrix notation,

$$\rho_k \mathbf{x}_k = \mathbf{D}_r \mathbf{P} \mathbf{x}_k \tag{7.9}$$

$$\rho_k \mathbf{y}_k = \mathbf{D}_c \mathbf{P}^T \mathbf{y}_k \,. \tag{7.10}$$

For Eqs. (7.3) and (7.4), ρ_k is the correlation between x_{ik} and y_{jk} such that

$$\begin{aligned} \rho_k &= \sum_{i=1}^{m} \sum_{j=1}^{n} p_{ij} \left(\frac{x_{ik} - \mathrm{E}(x_{ik})}{\mathrm{Var}(x_{ik})} \right) \left(\frac{y_{jk} - \mathrm{E}\left(y_{jk}\right)}{\mathrm{Var}\left(y_{jk}\right)} \right) \\ &= \sum_{i=1}^{m} \sum_{j=1}^{n} p_{ij} x_{ik} y_{jk} \,. \end{aligned} \tag{7.11}$$

In fact, as we shall describe in Sect. 7.4, this correlation is the maximum correlation between the row score x_{ik} and the column score y_{jk}; see also De Leeuw (1983).

To determine the set of row scores and column scores, $\mathbf{x}_k = (x_{1k}, x_{2k}, \ldots, x_{mk})^T$ and $\mathbf{y}_k = (y_{1k}, y_{2k}, \ldots, y_{nk})^T$, respectively, the analyst can make use of an iterative algorithm, or more simply (for those who have an appropriate function in their computing package) by applying an eigendecomposition to a square matrix resulting from these profiles. We shall now examine this iterative procedure and eigendecomposition in the next two sections.

7.3.3 The Iterative Approach

Suppose we confine our attention to determining the elements of $\mathbf{x}_k$ and $\mathbf{y}_k$ for $k = 1$ using reciprocal averaging. There are a few ways in which these two sets of scores can be iteratively determined. Here, we describe two such algorithms for calculating $\mathbf{x}_1, \mathbf{y}_1$ and ρ_1.

7.3.3.1 Convergence of the Row and Column Scores

The first algorithm we describe for determining $\mathbf{x}_1$, $\mathbf{y}_1$ and ρ_1 is based on the iterative process described by Nishisato (1980). The central feature of this algorithm is that it is performed until the convergence criterion is satisfied for each element of $\mathbf{x}_1$ and $\mathbf{y}_1$.

Define an initial set of values of $\mathbf{x}_1$ and denote this set by $\mathbf{x}_1^{(0)}$. It is then centred at zero and standardized so that

$$\mathbf{x}_1^{(0)} \Rightarrow \frac{\mathbf{x}_1^{(0)}}{\left(\mathbf{x}_1^{(0)}\right)^T \mathbf{D}_r \mathbf{x}_1^{(0)}} .$$

These row scores are then used to calculate the "initial" set of column scores (i.e., those values in the set $\mathbf{y}_1$) such that

$$\mathbf{y}_1^{(0)} = \mathbf{D}_r^{-1} \mathbf{P} \mathbf{x}_1^{(0)} .$$

These "initial" set of column scores is then centred at zero and standardized so that $\left(\mathbf{y}_1^{(0)}\right)^T \mathbf{D}_c \mathbf{y}_1^{(0)} = 1$. The standardization is achieved by

$$\mathbf{y}_1^{(0)} \Rightarrow \frac{\mathbf{y}_1^{(0)}}{\left(\mathbf{y}_1^{(0)}\right)^T \mathbf{D}_c \mathbf{y}_1^{(0)}}$$

and this set of scores is used to provide an update of the row scores, $\mathbf{x}_1^{(1)}$, at the next iteration so that

$$\mathbf{x}_1^{(1)} = \mathbf{D}_c^{-1} \mathbf{P}^T \mathbf{y}_1^{(0)} .$$

This set of scores is then centred at zero and standardized to ensure that $\left(\mathbf{x}_1^{(1)}\right)^T \mathbf{D}_r \mathbf{x}_1^{(1)} = 1$.

This iterative process continues the updating process so that, at the $l + 1$th iteration,

$$\mathbf{x}_1^{(l+1)} = \mathbf{D}_c^{-1} \mathbf{P}^T \mathbf{y}_1^{(l)} \quad \Rightarrow \quad \mathbf{y}_1^{(l+1)} = \mathbf{D}_r^{-1} \mathbf{P} \mathbf{x}_1^{(l+1)} .$$

The stopping criteria applied to this algorithm is defined so that

$$\left| \mathbf{x}_1^{(l+1)} - \mathbf{x}_1^{(l)} \right| < \epsilon$$

and

$$\left| \mathbf{y}_1^{(l+1)} - \mathbf{y}_1^{(l)} \right| < \epsilon$$

where ϵ is some small value, say $\epsilon = 0.0001$. When convergence is achieved, $\mathbf{x}_1^{(l+1)}$ and $\mathbf{y}_1^{(l+1)}$ are deemed to be the vector of row and column scores with a correlation of

$$\rho_1 = \sum_{i=1}^{m} \sum_{j=1}^{n} p_{ij} x_{i1}^{(l+1)} y_{j1}^{(l+1)} .$$

One may see that this iterative process is executed until convergence is achieved for each of the m values of $\mathbf{x}_1$ and for each of the n values of $\mathbf{y}_1$. This is potentially problematic if m and/or n is very large since the convergence criteria are applied to $m + n$ values. This process also means that $\mathbf{x}_1$ and $\mathbf{y}_1$ at each iteration is centred as an additional step of the algorithm rather than being built into each updated calculation of the scores.

7.3.3.2 Convergence of the Correlation

A second, and potentially more efficient, iterative process for determining the row scores in $\mathbf{x}_1$, the column scores in $\mathbf{y}_1$ and the correlation between them (ρ_1) is to adopt the following strategy.

As Nishisato (1980) did, define an initial set of row scores, say, and denote this set by $\mathbf{x}_1^{(0)}$. It is then standardized so that $\left(\mathbf{x}_1^{(0)}\right)^T \mathbf{D}_r \mathbf{x}_1^{(0)} = 1$. Note that here these scores are not yet centred at zero. This is not a problem since the centring is carried out at the next phase of the algorithm that updates the values in $\mathbf{x}_1$. An initial value of ρ_1 is defined; a good starting value is $\rho_1^{(0)} = 1$. This initial value of the correlation, and the initial set of row scores, is then used to calculate the "initial" set of column scores so that

$$\mathbf{y}_1^{(0)} = \frac{1}{\rho_1^{(0)}} \left(\mathbf{D}_r^{-1}\mathbf{P} - \mathbf{1}_r \mathbf{c}^T\right) \mathbf{x}_1^{(0)} . \tag{7.12}$$

Note that here the initial set of column scores is centred at zero within this iterative process. They can then be standardized so that $\left(\mathbf{y}_1^{(0)}\right)^T \mathbf{D}_c \mathbf{y}_1^{(0)} = 1$. An updated value of the correlation coefficient, ρ_1, is then calculated by

$$\rho_1^{(1)} = \sum_{i=1}^{m} \sum_{j=1}^{n} p_{ij} x_{i1}^{(0)} y_{j1}^{(0)} .$$

The values of $\mathbf{x}_1$, $\mathbf{y}_1$ and ρ_1 are then updated as follows. The values of $\mathbf{x}_1$ are calculated at the next iteration such that

$$\mathbf{x}_1^{(1)} = \frac{1}{\rho_1^{(1)}} \left(\mathbf{D}_r^{-1}\mathbf{P} - \mathbf{1}_r \mathbf{c}^T\right) \mathbf{y}_1^{(0)} .$$

Note that here the centring of the $\mathbf{x}_1$ values at zero is incorporated into this step. One only then needs to ensure that they are standardized so that $\left(\mathbf{x}_1^{(1)}\right)^T \mathbf{D}_r \mathbf{x}_1^{(1)} = 1$. An update of the set of column scores, $\mathbf{y}_1^{(1)}$, is then calculated from

$$\mathbf{y}_1^{(1)} = \frac{1}{\rho_1^{(1)}} \left(\mathbf{D}_c^{-1}\mathbf{P}^T - \mathbf{1}_c\mathbf{r}^T\right) \mathbf{x}_1^{(1)}$$

and are then standardized so that $\left(\mathbf{y}_1^{(1)}\right) \mathbf{D}_c \mathbf{y}_1^{(1)} = 1$. The correlation coefficient is then updated by

$$\rho_1^{(2)} = \sum_{i=1}^{m} \sum_{j=1}^{n} p_{ij} x_{i1}^{(1)} y_{j1}^{(1)}.$$

These steps are repeatedly performed so that, at the lth iteration,

$$\mathbf{x}_1^{(l)} = \frac{1}{\rho_1^{(l-1)}} \left(\mathbf{D}_r^{-1}\mathbf{P} - \mathbf{1}_r\mathbf{c}^T\right) \mathbf{y}_1^{(l-1)}, \tag{7.13}$$

$$\mathbf{y}_1^{(l)} = \frac{1}{\rho_1^{(l)}} \left(\mathbf{D}_c^{-1}\mathbf{P}^T - \mathbf{1}_c\mathbf{r}^T\right) \mathbf{x}_1^{(l)} \tag{7.14}$$

where

$$\rho_1^{(l)} = \sum_{i=1}^{m} \sum_{j=1}^{n} p_{ij} x_{i1}^{(l-1)} y_{j1}^{(l-1)}.$$

The correlation between $\mathbf{x}_1^{(l)}$ and $\mathbf{y}_1^{(l)}$ is then determined by

$$\rho_1^{(l+1)} = \sum_{i=1}^{m} \sum_{j=1}^{n} p_{ij} x_{i1}^{(l)} y_{j1}^{(l)}$$

and a stopping criterion is applied only to this correlation coefficient. It is defined such that if $\left|\rho_1^{(l+1)} - \rho_1^{(l)}\right| < \epsilon$ is satisfied, for some small value of ϵ, then $\mathbf{x}_1^{(l)}$ and $\mathbf{y}_1^{(l)}$ are set of row and column scores where the correlation between them is $\rho_1^{(l+1)}$.

7.3.3.3 Remarks on These Algorithms

Practical applications of the implementation of these two algorithms show that they produce the same set of row scores $\mathbf{x}_1$. They also produce identical column scores $\mathbf{y}_1$ while the correlation between these scores, ρ_1 is also identical for the two algorithms.

However, the stopping criterion of the first algorithm applies to the m values of $\mathbf{x}_1$ and the n values of $\mathbf{y}_1$. Therefore, the analyst must ensure that convergence to an appropriate level is achieved for $m+n$ values. This is likely to require more iterations for convergence to be satisfied than the second algorithm which applies the stopping criterion to only a single value, ρ_1. The second algorithm also ensures that the row and column scores are centred at zero at each update of their values, rather than performing the centring as an additional step after each update of the scores.

7.3.4 The Role of Eigendecomposition

An alternative approach to determining the row and column scores of a two-way contingency table, and the correlation between them, is to use eigendecomposition. Doing so produces the same solution to $\mathbf{x}_1$, $\mathbf{y}_1$ and ρ_1 as the above algorithms, but does so directly. The more direct nature of their calculation can be easily undertaken in R, for example, using the `eigen()` function. The advantage of determining these scores (and their correlation) in this manner is that we need to not be confined to calculating their values for a single dimension. Here, we shall consider the more general case of determining $\mathbf{x}_k$, $\mathbf{y}_k$ and ρ_k for $k = 1, 2, \ldots, K$ where $K = \min(m, n) - 1$ and

$$\mathbf{x}_k^T \mathbf{D}_r \mathbf{x}_k = 1 \quad \text{and} \quad \mathbf{y}_k^T \mathbf{D}_c \mathbf{y}_k = 1 .$$

The first thing to note is that Eqs. (7.3) and (7.4) can be expressed in matrix form by

$$\rho_k \mathbf{x}_k = \left(\mathbf{D}_r^{-1}\mathbf{P} - \mathbf{1}_r \mathbf{c}^T\right) \mathbf{y}_k \tag{7.15}$$

$$\rho_k \mathbf{y}_k = \left(\mathbf{D}_c^{-1}\mathbf{P}^T - \mathbf{1}_c \mathbf{r}^T\right) \mathbf{x}_k , \tag{7.16}$$

respectively. One may note the similarities of these two equations with those of Hirschfeld (1935) [Eq. 2]. While Eqs. (7.15) and (7.16) incorporate the centring of the row and column scores through the specification of $\mathbf{x}_k$ and $\mathbf{y}_k$ (which are defined to have a zero weighted mean and unit variance), Hirschefeld's equations imply that his scores were not centred, but that the centring was incorporated by subtracting from his scores the weighted mean of those scores.

One may also observe that, using the properties given by Eqs. (7.7) and (7.8), then Eqs. (7.15) and (7.16) may be alternatively expressed as

$$\mathbf{x}_k = \frac{1}{\rho_k}\mathbf{D}_r^{-1}\mathbf{P}\mathbf{y}_k$$

$$\mathbf{y}_k = \frac{1}{\rho_k}\mathbf{D}_c^{-1}\mathbf{P}^T\mathbf{x}_k$$

which are just a simple rearrangement of Eqs. (7.9) and (7.10) and are referred to as *formules barycentriques* (Benzécri 1973), *dual relations* or *duality* (Nishisato 1980,

p. 59). Note that these are akin to the uncentred versions of Eqs. (7.13) and (7.14) used to iteratively determine the row and column scores using the algorithm described in Sect. 7.3.3.2.

Pre-multiplying both sides of Eq. (7.15) by $\rho_k \mathbf{D}_r^{1/2}$ gives

$$\begin{aligned} \rho_k^2 \left(\mathbf{D}_r^{1/2}\mathbf{x}_k\right) &= \rho_k \mathbf{D}_r^{1/2}\left(\mathbf{D}_r^{-1}\mathbf{P} - \mathbf{1}_r\mathbf{c}^T\right)\mathbf{y}_k \\ &= \left[\mathbf{D}_r^{-1/2}\left(\mathbf{P} - \mathbf{r}\mathbf{c}^T\right)\mathbf{D}_c^{-1/2}\right]\left(\rho_k \mathbf{D}_c^{1/2}\mathbf{y}_k\right) \end{aligned} \quad (7.17)$$

while pre-multiplying Eq. (7.16) by $\mathbf{D}_c^{1/2}$ gives

$$\rho_k \left(\mathbf{D}_c^{1/2}\mathbf{y}_k\right) = \left[\mathbf{D}_c^{-1/2}\left(\mathbf{P} - \mathbf{r}\mathbf{c}^T\right)^T \mathbf{D}_r^{-1/2}\right]\left(\mathbf{D}_r^{1/2}\mathbf{x}_k\right) . \quad (7.18)$$

By substituting Eqs. (7.18) into (7.17) gives

$$\rho_k^2 \left(\mathbf{D}_r^{1/2}\mathbf{x}_k\right) = \left[\mathbf{D}_r^{-1/2}\left(\mathbf{P} - \mathbf{r}\mathbf{c}^T\right)\mathbf{D}_c^{-1/2}\right]\left[\mathbf{D}_c^{-1/2}\left(\mathbf{P} - \mathbf{r}\mathbf{c}^T\right)^T \mathbf{D}_r^{-1/2}\right]\left(\mathbf{D}_r^{1/2}\mathbf{x}_k\right) . \quad (7.19)$$

To help simplify Eq. (7.19), let

$$\mathbf{Z} = \mathbf{D}_r^{-1/2}\left(\mathbf{P} - \mathbf{r}\mathbf{c}^T\right)\mathbf{D}_c^{-1/2} . \quad (7.20)$$

By doing so we can see that the $(i,\ j)$th element of $\mathbf{Z}$ is

$$Z_{ij} = \frac{p_{ij} - p_{i\bullet}p_{\bullet j}}{\sqrt{p_{i\bullet}p_{\bullet j}}} \quad (7.21)$$

which is just the normalization of the $(i,\ j)$th cell proportion under the hypothesis of independence when the cell frequencies of the contingency table are assumed to be random variables from a Poisson distribution. We refer to this quantity as the $(i,\ j)$th *Pearson residual* since Pearson's chi-squared is just f_t times the sum-of-squares of the Z_{ij} values. Note also that this value can be expressed as the weighted centred row profile and as the weighted centred column profile since

$$Z_{ij} = \sqrt{\frac{p_{i\bullet}}{p_{\bullet j}}}\left(\frac{p_{ij}}{p_{i\bullet}} - p_{\bullet j}\right) = \sqrt{\frac{p_{\bullet j}}{p_{i\bullet}}}\left(\frac{p_{ij}}{p_{\bullet j}} - p_{i\bullet}\right) .$$

By using the definition of $\mathbf{Z}$ given by Eq. (7.20), then Eq. (7.19) simplifies to

$$\rho_k^2 \left(\mathbf{D}_r^{1/2}\mathbf{x}_k\right) = \mathbf{Z}\mathbf{Z}^T \left(\mathbf{D}_r^{1/2}\mathbf{x}_k\right)$$

which, upon rearranging, becomes

$$\left(\mathbf{Z}\mathbf{Z}^T - \rho_k^2 \mathbf{I}_m\right)\left(\mathbf{D}_r^{1/2}\mathbf{x}_k\right) = \mathbf{0}_m \quad (7.22)$$

where $\mathbf{I}_m$ is a $m \times m$ identity matrix and $\mathbf{0}_m$ is a vector of zeros of length m.

Suppose we rescale $\mathbf{x}_k$ so that $\tilde{\mathbf{x}}_k = \mathbf{D}_r^{1/2}\mathbf{x}_k$. Then $\tilde{\mathbf{x}}_k^T\tilde{\mathbf{x}}_k = 1$ while Eq. (7.22) can be rewritten as

$$\left(\mathbf{Z}\mathbf{Z}^T - \rho_k^2\mathbf{I}_m\right)\tilde{\mathbf{x}}_k = \mathbf{0}_m . \tag{7.23}$$

Here, ρ_k^2 is the kth eigenvalue of the diagonal matrix $\mathbf{Z}\mathbf{Z}^T$ and $\tilde{\mathbf{x}}_k$ is the kth eigenvector. We can also show that the set of column score, $\mathbf{y}_k$, can be determined by solving

$$\left(\mathbf{Z}^T\mathbf{Z} - \rho_k^2\mathbf{I}_n\right)\tilde{\mathbf{y}}_k = \mathbf{0}_n \tag{7.24}$$

where $\tilde{\mathbf{y}}_k = \mathbf{D}_c^{1/2}\mathbf{y}_k$ so that $\tilde{\mathbf{y}}_k^T\tilde{\mathbf{y}}_k = 1$. In this case, $\mathbf{I}_n$ is a $n \times n$ identity matrix and $\mathbf{0}_n$ is a zero vector of length n. Therefore, the row scores x_{ik} (or their weighted version $\tilde{x}_{ik}$), can be obtained via Eq. (7.24) which is just the eigendecomposition of $\mathbf{Z}^T\mathbf{Z}$ so that ρ_k^2 is the kth largest eigenvalue of this matrix.

7.3.5 The Role of Singular Value Decomposition

The previous section demonstrated that to determine the row scores and columns scores contained in $\mathbf{x}_k$ and $\mathbf{y}_k$, respectively, two eigendecompositions can be performed; one on the matrix $\mathbf{Z}^T\mathbf{Z}$ and another on the matrix $\mathbf{Z}\mathbf{Z}^T$. Rather than carrying out this method of decomposition twice, the scores can be calculated using a single approach—*singular value decomposition* (SVD). This method of decomposition can be applied to

$$\mathbf{Z} = \mathbf{D}_r^{-1/2}\left(\mathbf{P} - \mathbf{r}\mathbf{c}^T\right)\mathbf{D}_c^{-1/2} \tag{7.25}$$

such that

$$\mathbf{Z} = \tilde{\mathbf{X}}\Lambda\tilde{\mathbf{Y}}^T \tag{7.26}$$

where $\tilde{\mathbf{X}} = (\tilde{\mathbf{x}}_1 \;\; \tilde{\mathbf{x}}_2 \ldots \tilde{\mathbf{x}}_K)$ and $\tilde{\mathbf{Y}} = (\tilde{\mathbf{y}}_1 \;\; \tilde{\mathbf{y}}_2 \ldots \tilde{\mathbf{y}}_K)$ for $K = \min(m, n) - 1$. These matrices are of size $m \times K$ and $n \times K$, respectively, and are normalized so that

$$\tilde{\mathbf{X}}^T\tilde{\mathbf{X}} = \mathbf{I}_K \quad \text{and} \quad \tilde{\mathbf{Y}}^T\tilde{\mathbf{Y}} = \mathbf{I}_K . \tag{7.27}$$

The diagonal matrix Λ is of size $K \times K$ and contains the square root of the K eigenvalues that are referred to as *singular values* and are all positive and arranged in descending order so that

$$\rho_1 > \rho_2 > \cdots > \rho_K .$$

The set of row and column scores, $\tilde{\mathbf{x}}_k$ and $\tilde{\mathbf{y}}_k$, and ρ_k can be easily calculated using singular value decomposition using the `svd()` function in `R` which imposes the property on the matrix of scores defined by Eq. (7.27). The set of row scores, $\mathbf{x}_k$, and column scores, $\mathbf{y}_k$ can also be easily calculated by $\mathbf{x}_k = \mathbf{D}_r^{-1/2}\tilde{\mathbf{x}}_k$ and $\mathbf{y}_k = \mathbf{D}_c^{-1/2}\tilde{\mathbf{y}}_k$.

7.3.6 Models of Correlation and Association

The advantage of viewing the scoring of the rows and columns from this perspective is that the resulting row and column scores have clear links to association, correlation, and correspondence models. To show this since, from Eqs. (7.25) and (7.26), we obtain

$$\mathbf{D}_r^{-1/2}\left(\mathbf{P} - \mathbf{r}\mathbf{c}^T\right)\mathbf{D}_c^{-1/2} = \tilde{\mathbf{X}}\Lambda\tilde{\mathbf{Y}}^T .$$

Rearranging this expression leads to the following results

$$\begin{aligned}\mathbf{P} &= \mathbf{r}\mathbf{c}^T\left(1 + \left(\mathbf{D}_r^{-1/2}\tilde{\mathbf{X}}\right)\Lambda\left(\mathbf{D}_c^{-1/2}\tilde{\mathbf{Y}}\right)^T\right)\\ &= \mathbf{r}\mathbf{c}^T\left(1 + \mathbf{X}\Lambda\mathbf{Y}^T\right)\end{aligned} \tag{7.28}$$

where $\mathbf{X} = \mathbf{D}_r^{-1/2}\tilde{\mathbf{X}}$ and $\mathbf{Y} = \mathbf{D}_c^{-1/2}\tilde{\mathbf{Y}}$ such that

$$\mathbf{X} = \left(\mathbf{x}_1 \quad \mathbf{x}_2 \quad \ldots \quad \mathbf{x}_K\right)$$

and

$$\mathbf{Y} = \left(\mathbf{y}_1 \quad \mathbf{y}_1 \quad \ldots \quad \mathbf{y}_K\right) .$$

Therefore, Eq. (7.28) can be expressed in these terms as

$$\mathbf{P} = \mathbf{r}\mathbf{c}^T\left(1 + \mathbf{x}_1\rho_1\mathbf{y}_1^T + \mathbf{x}_2\rho_2\mathbf{y}_2^T + \cdots + \mathbf{x}_K\rho_K\mathbf{y}_K^T\right) \tag{7.29}$$

so that the $(i,\ j)$th cell frequency of the two-way contingency table can be calculated by

$$f_{ij} = \frac{f_{i\bullet}f_{\bullet j}}{f_t}\left(1 + \mathbf{x}_1\rho_1\mathbf{y}_1^T + \mathbf{x}_2\rho_2\mathbf{y}_2^T + \cdots + \mathbf{x}_K\rho_K\mathbf{y}_K^T\right) .$$

Equation (7.29) shows that when there is complete independence between the row and column variables (so that $\rho_k = 0$ for $k = 1, \ldots, K$) then $\mathbf{P} = \mathbf{r}\mathbf{c}^T$; these are the expected values of the cell proportions under complete independence between the two categorical variables of the contingency table. This result also shows that one may reconstitute (or approximate) the cell frequencies of $\mathbf{P}$ by just using the one-dimensional solution of $\mathbf{x}_1, \mathbf{y}_1$ and ρ_1 from the reciprocal averaging of the contingency table. In doing so, the matrix of cell proportions, $\mathbf{P}$ can be approximated by

$$\mathbf{P} \approx \mathbf{r}\mathbf{c}^T\left(1 + \mathbf{x}_1\rho_1\mathbf{y}_1^T\right) . \tag{7.30}$$

A problem with doing this result is that, depending on how big, or small, K is, the approximation can be quite bad. Worse still, it is possible for it to produce estimates of the joint proportions (and frequencies) that are negative! One way to overcome this issue is to note that a first order Maclaurin series expansion of the exponential function is

$$\exp(x) = 1 + x$$

for all x. Therefore, $\mathbf{P}$ may be approximated using the exponential model

$$\mathbf{P} \approx \mathbf{rc}^T \exp\left(\mathbf{x}_1 \rho_1 \mathbf{y}_1^T\right) \tag{7.31}$$

and ensures that approximations of p_{ij} are non-negative. More generally, this exponential form of the model may be applied for all k, $k = 1, \ldots, K$ so that Eq. (7.29) can be approximated by

$$\mathbf{P} = \mathbf{rc}^T \exp\left(\mathbf{x}_1 \rho_1 \mathbf{y}_1^T + \mathbf{x}_2 \rho_2 \mathbf{y}_2^T + \cdots + \mathbf{x}_K \rho_K \mathbf{y}_K^T\right) .$$

However, unlike Eq. (7.29), this result will not produce perfectly reconstituted values of $\mathbf{P}$.

7.4 Canonical Correlation Analysis

7.4.1 An Overview

In Sect. 7.3, we demonstrated how scores for each of the m rows and n columns of a two-way contingency table can be obtained via an eigendecomposition. From this discussion, we showed that the square root of the kth largest eigenvalue of $\mathbf{Z}^T\mathbf{Z}$ and $\mathbf{ZZ}^T$ is the correlation between the set of row scores (where the ith score is denoted by x_{ik}) and the set of column scores (where the jth score is denoted by y_{jk}). Here, we show using canonical correlation analysis that this correlation is the maximum possible correlation between these scores and that this approach also reduces to the same set of row and column scores that reciprocal averaging yields. From a historical perspective, the original idea of canonical correlation analysis comes from Hirschfeld (1935) who examined the idea of calculating scores for a multi-way contingency table. One may also refer to, for example, Gower et al. (2011, Sect. 7.2.5) for a more recent discussion of this analysis.

7.4.2 The Method

Following on from our discussion in Sect. 7.3 we define, using matrix notation, the correlation between the (weighted) set of row scores and column scores by

$$\begin{aligned}
\rho_k &= \mathrm{Corr}\left(\mathbf{x}_k,\ \mathbf{y}_k\right) \\
&= \frac{\left(\mathbf{x}_k - \mathrm{E}\left(\mathbf{x}_k\right)\right)^T}{\sqrt{\mathrm{Var}\left(\mathbf{x}_k - \mathrm{E}\left(\mathbf{x}_k\right)\right)}}\mathbf{P}\frac{\left(\mathbf{y}_k - \mathrm{E}\left(\mathbf{y}_k\right)\right)}{\sqrt{\mathrm{Var}\left(\mathbf{y}_k - \mathrm{E}\left(\mathbf{y}_k\right)\right)}} \\
&= \frac{\mathbf{x}_k^T\mathbf{P}\mathbf{y}_k}{\sqrt{\left(\mathbf{x}_k^T\mathbf{D}_r\mathbf{x}_k\right)\left(\mathbf{y}_k^T\mathbf{D}_c\mathbf{y}_k\right)}}\,.
\end{aligned}$$

Note that here we are not yet imposing any particular property on $\mathbf{x}_k$ or $\mathbf{y}_k$, although by virtue of their vector form then $\mathbf{x}_k^T\mathbf{D}_r\mathbf{x}_k$ and $\mathbf{y}_k^T\mathbf{D}_c\mathbf{y}_k$ must be constants. Squaring this correlation gives

$$\begin{aligned}
\rho_k^2 &= \frac{\left(\mathbf{x}_k^T\mathbf{P}\mathbf{y}_k\right)^2}{\left(\mathbf{x}_k^T\mathbf{D}_r\mathbf{x}_k\right)\left(\mathbf{y}_k^T\mathbf{D}_c\mathbf{y}_k\right)} \\
&= \left(\mathbf{x}_k^T\mathbf{D}_r\mathbf{x}_k\right)^{-1}\left(\mathbf{x}_k^T\mathbf{P}\mathbf{y}_k\right)^2\left(\mathbf{y}_k^T\mathbf{D}_c\mathbf{y}_k\right)^{-1}\,. \qquad (7.32)
\end{aligned}$$

To determine the expression that yields the maximum value of this (squared) correlation, and the conditions under which it exists, we re-express ρ_k^2 by grouping together the $\mathbf{x}_k$ terms so that

$$\rho_k^2 = \left[\left(\mathbf{x}_k^T\mathbf{D}_r\mathbf{x}_k\right)^{-1}\left(\mathbf{x}_k^T\mathbf{P}\mathbf{y}_k\right)^2\right]\left(\mathbf{y}_k^T\mathbf{D}_c\mathbf{y}_k\right)^{-1}\,.$$

We therefore maximize this squared correlation by differentiating it with respect to $\mathbf{x}_k$ so that

$$\begin{aligned}
\frac{\partial}{\partial\mathbf{x}_k}\rho_k^2 &= \left[\frac{\partial}{\partial\mathbf{x}_k}\left(\mathbf{x}_k^T\mathbf{D}_r\mathbf{x}_k\right)^{-1}\left(\mathbf{x}_k^T\mathbf{P}\mathbf{y}_k\right)^2\right]\left(\mathbf{y}_k^T\mathbf{D}_c\mathbf{y}_k\right)^{-1} \\
&= \left\{-\left(\mathbf{x}_k^T\mathbf{D}_r\mathbf{x}_k\right)^{-2}\left(2\mathbf{D}_r\mathbf{x}_k\right)\left(\mathbf{x}_k^T\mathbf{P}\mathbf{y}_k\right)^2 + \right. \\
&\qquad \left. 2\left(\mathbf{x}_k^T\mathbf{D}_r\mathbf{x}_k\right)^{-1}\left(\mathbf{x}_k^T\mathbf{P}\mathbf{y}_k\right)\left(\mathbf{P}\mathbf{x}_k\right)\right\}\left(\mathbf{y}_k^T\mathbf{D}_c\mathbf{y}_k\right)^{-1} \\
&= 2\left(\mathbf{x}_k^T\mathbf{D}_r\mathbf{x}_k\right)^{-1}\left(\mathbf{x}_k^T\mathbf{P}\mathbf{y}_k\right)\left(\mathbf{P}\mathbf{x}_k\right)\left(\mathbf{y}_k^T\mathbf{D}_c\mathbf{y}_k\right)^{-1} - \\
&\qquad 2\left(\mathbf{x}_k^T\mathbf{D}_r\mathbf{x}_k\right)^{-2}\left(\mathbf{D}_r\mathbf{x}_k\right)\left(\mathbf{x}_k^T\mathbf{P}\mathbf{y}_k\right)^2\left(\mathbf{y}_k^T\mathbf{D}_c\mathbf{y}_k\right)^{-1} \\
&= 0\,. \qquad (7.33)
\end{aligned}$$

Similarly, with respect to $\mathbf{y}_k$, we set the derivative of the squared correlation to zero such that

$$\begin{aligned}\frac{\partial}{\partial \mathbf{y}_k}\rho_k^2 &= 2\left(\mathbf{y}_k^T\mathbf{D}_c\mathbf{y}_k\right)^{-1}\left(\mathbf{y}_k^T\mathbf{P}\mathbf{y}_k\right)\left(\mathbf{P}^T\mathbf{x}_k\right)\left(\mathbf{x}_k^T\mathbf{D}_r\mathbf{x}_k\right)^{-1} - \\ &\qquad 2\left(\mathbf{y}_k^T\mathbf{D}_c\mathbf{y}_k\right)^{-2}\left(\mathbf{D}_c\mathbf{y}_k\right)\left(\mathbf{y}_k^T\mathbf{P}^T\mathbf{x}_k\right)^2\left(\mathbf{x}_k^T\mathbf{D}_r\mathbf{x}_k\right)^{-1} \\ &= 0\,. \end{aligned} \tag{7.34}$$

To simplify these results, we now introduce that $\mathbf{x}_k$ and $\mathbf{y}_k$ have the property $\mathbf{x}_k^T\mathbf{D}_r\mathbf{x}_k = 1$ and $\mathbf{y}_k^T\mathbf{D}_c\mathbf{y}_k = 1$, respectively. By doing so, Eqs. (7.33) and (7.34) reduce to

$$\frac{\partial}{\partial \mathbf{x}_k}\rho_k^2 = 2\rho_k\left(\mathbf{P}\mathbf{y}_k\right) - 2\rho_k^2\left(\mathbf{D}_r\mathbf{x}_k\right) = 0 \tag{7.35}$$

$$\frac{\partial}{\partial \mathbf{y}_k}\rho_k^2 = 2\rho_k\left(\mathbf{x}_k^T\mathbf{P}\right) - 2\rho_k^2\left(\mathbf{y}_k^T\mathbf{D}_c\right) = 0\,, \tag{7.36}$$

respectively. Therefore, Eq. (7.32) simplifies to

$$\rho_k^2 = \left(\mathbf{x}_k^T\mathbf{P}\mathbf{y}_k\right)^2$$

or, elementwise,

$$\rho_k = \sum_{i=1}^{m}\sum_{j=1}^{n} p_{ij}x_{ik}y_{jk}$$

which is just Eq. (7.11). We can verify that this is indeed the maximum (squared) correlation between the set of row scores $\mathbf{x}_k$ and column scores $\mathbf{y}_k$ since

$$\frac{\partial^2}{\partial \mathbf{x}_k^2}\rho_k^2 = -2\rho_k^2\mathbf{D}_r < 0$$

$$\frac{\partial^2}{\partial \mathbf{y}_k^2}\rho_k^2 = -2\rho_k^2\mathbf{D}_c < 0$$

for all elements of $\mathbf{x}_k$ and $\mathbf{y}_k$ and values of ρ_k.

To verify that the solution to $\mathbf{x}_k$ and $\mathbf{y}_k$ is exactly those obtained via reciprocal averaging, Eqs. (7.35) and (7.36) reduce to

$$\left(\mathbf{P}\mathbf{y}_k\right) = \rho_k\left(\mathbf{D}_r\mathbf{x}_k\right)$$

$$\left(\mathbf{x}_k^T\mathbf{P}\right) = \rho_k\left(\mathbf{y}_k^T\mathbf{D}_c\right)$$

which can be alternatively, and equivalently, expressed as

$$\rho_k\mathbf{x}_k = \left(\mathbf{D}_r^{-1}\mathbf{P}\right)\mathbf{y}_k = \left(\mathbf{D}_r^{-1}\mathbf{P} - \mathbf{1}_r\mathbf{c}^T\right)\mathbf{y}_k$$

$$\rho_k\mathbf{y}_k = \left(\mathbf{D}_c^{-1}\mathbf{P}^T\right)\mathbf{x}_k = \left(\mathbf{D}_c^{-1}\mathbf{P}^T - \mathbf{1}_c\mathbf{r}^T\right)\mathbf{x}_k$$

since $\mathbf{r}^T \mathbf{x}_k = 0$ and $\mathbf{c}^T \mathbf{y}_k = 0$. These equations are just those of Eqs. (7.15) and (7.16), respectively. Therefore, canonical correlation analysis yields row and column scores, $\mathbf{x}_k$ and $\mathbf{y}_k$, respectively, that are identical to those obtained via reciprocal averaging with ρ_k being the maximum possible correlation between them.

Nishisato (1980) [p. 28] considered the maximization of a similar quantity to Eq. (7.32) but did so by incorporating a Lagrangian multiplier to reflect the properties that are eventually given to $\mathbf{x}_k$ and $\mathbf{y}_k$. The careful reader will note that we have considered that $\mathbf{x}_k$ and $\mathbf{y}_k$ have the property defined by Eqs. (7.5) and (7.6), respectively. However, many in the scaling literature refer to them as being "constraints". We have avoided this term instead of deciding to keep in mind that Gower (1989) [p. 222] noted such constraints are "irrelevant" and that "these settings are conveniences that should not be regarded as constraints". Therefore, there is no reason why alternative properties cannot be considered for $\mathbf{x}_k$ and $\mathbf{y}_k$.

7.5 Example

7.5.1 One-Dimensional Solution via Reciprocal Averaging

Suppose we consider the 5×5 contingency table of Table 3.1 which is reproduced here as Table 7.1. For this data, the row and column marginal proportions are

$$\mathbf{r} = (0.2 \quad 0.2 \quad 0.2 \quad 0.2 \quad 0.2)^T$$

and

$$\mathbf{c} = (0.1929 \quad 0.3357 \quad 0.1714 \quad 0.0929 \quad 0.2071)^T ,$$

respectively. From this table, we also find that the matrix of joint proportions is

$$\mathbf{P} = \begin{pmatrix} 0.1071 & 0.0571 & 0.0214 & 0.0143 & 0.0000 \\ 0.0357 & 0.1214 & 0.0286 & 0.0000 & 0.0143 \\ 0.0429 & 0.0929 & 0.0286 & 0.0214 & 0.0143 \\ 0.0000 & 0.0500 & 0.0500 & 0.0357 & 0.0643 \\ 0.0071 & 0.0143 & 0.0429 & 0.0214 & 0.1143 \end{pmatrix} . \quad (7.37)$$

We also note that the matrix of Pearson residuals for Table 7.1 is

$$\mathbf{Z} = \begin{pmatrix} 0.3491 & -0.0386 & -0.0694 & -0.0314 & -0.2035 \\ -0.0145 & 0.2095 & -0.0309 & -0.1363 & -0.1334 \\ 0.0218 & 0.0992 & -0.0309 & 0.0210 & -0.1334 \\ -0.1964 & -0.0662 & 0.0849 & 0.1258 & 0.1123 \\ -0.1600 & -0.2040 & 0.0463 & 0.0210 & 0.3579 \end{pmatrix} . \quad (7.38)$$

Table 7.1 Contingency table of *Propensity* to take sleeping pills and frequency of *Nightmares*

Propensity	Nightmares					
	Never	Rarely	Sometimes	Often	Always	Total
Strongly against	15	8	3	2	0	28
Against	5	17	4	0	2	28
Neutral	6	13	4	3	2	28
For	0	7	7	5	9	28
Strongly For	1	2	6	3	16	28
Total	27	47	24	13	29	140

We can use this information to determine the row and column scores as follows. Firstly, we define an initial set of row scores. Suppose they are defined to be

$$\mathbf{x}_1^{(0)} = (1,\ 2,\ 3,\ 4,\ 5)^T$$

although any set of initial scores can be chosen. Therefore, for this choice of row scores, we find that

$$\left(\mathbf{x}_1^{(0)}\right)^T \mathbf{D}_r \mathbf{x}_1^{(0)} = 11\,.$$

So, the set of initial row scores are standardized so that

$$\begin{aligned}\mathbf{x}_1^{(0)} &= \frac{(1,\ 2,\ 3,\ 4,\ 5)^T}{\sqrt{11}} \\ &= (0.3015,\ 0.6030,\ 0.9045\ 1.2060,\ 1.5076)\ .\end{aligned}$$

These scores can now be used to determine the initial set column scores, denoted by $\mathbf{y}_1^{(0)}$. Using Eq. (7.12), we get the set of non-normalized column scores

$$
\begin{aligned}
\mathbf{y}_1^{(0)} &= \frac{1}{\rho_1^{(0)}}\left(\mathbf{D}_r^{-1}\mathbf{P} - \mathbf{1}_r\mathbf{c}^T\right)\left(\mathbf{x}_1^{(0)}\right) \\
&= \frac{1}{1}\begin{pmatrix} 0.3429 & -0.0500 & -0.0643 & -0.0214 & -0.2071 \\ -0.0143 & 0.2714 & -0.0286 & -0.0929 & -0.1357 \\ 0.0214 & 0.1286 & -0.0286 & 0.0143 & -0.1357 \\ -0.1929 & -0.0857 & 0.0786 & 0.0857 & 0.1143 \\ -0.1571 & -0.2643 & 0.0429 & 0.0143 & 0.3643 \end{pmatrix}\begin{pmatrix} 0.3015 \\ 0.6030 \\ 0.9045 \\ 1.2060 \\ 1.5076 \end{pmatrix} \\
&= \begin{pmatrix} -0.0285 \\ -0.0269 \\ -0.0250 \\ -0.0247 \\ -0.0229 \end{pmatrix}.
\end{aligned}
$$

From these column scores, we find that

$$
\left(\mathbf{y}_1^{(0)}\right)^T \mathbf{D}_c\mathbf{y}_1^{(0)} = 0.00067
$$

and so they can be standardized as follows

$$
\begin{aligned}
\mathbf{y}_1^{(0)} &= \frac{(1,\ 2,\ 3,\ 4,\ 5)^T}{\sqrt{0.00067}} \\
&= (-1.0988,\ -1.0362,\ -0.9661\ -0.9525,\ -0.8855)\ .
\end{aligned}
$$

With $\mathbf{x}_1^{(0)}$ and $\mathbf{y}_1^{(0)}$ now determined, the initial value of the correlation between the elements of these vectors is

$$
\begin{aligned}
\rho_1^{(0)} &= \left(\mathbf{x}_1^{(0)}\right)^T \mathbf{P}\mathbf{y}_1^{(0)} \\
&= \begin{pmatrix} 0.3015 \\ 0.6030 \\ 0.9045 \\ 1.2060 \\ 1.5076 \end{pmatrix}^T \begin{pmatrix} 0.1071 & 0.0571 & 0.0214 & 0.0143 & 0.0000 \\ 0.0357 & 0.1214 & 0.0286 & 0.0000 & 0.0143 \\ 0.0429 & 0.0929 & 0.0286 & 0.0214 & 0.0143 \\ 0.0000 & 0.0500 & 0.0500 & 0.0357 & 0.0643 \\ 0.0071 & 0.0143 & 0.0429 & 0.0214 & 0.1143 \end{pmatrix}\begin{pmatrix} -0.0285 \\ -0.0269 \\ -0.0250 \\ -0.0247 \\ -0.0229 \end{pmatrix} \\
&= -0.8823\,.
\end{aligned}
$$

These set of initial scores, $\mathbf{x}_1^{(0)}$ and $\mathbf{y}_1^{(0)}$, and the correlation between them, $\rho_1^{(0)}$, can then be used to update $\mathbf{x}_1$ and $\mathbf{y}_1$ using the algorithm outlined in Sect. 7.3.3.1 until convergence is achieved for these scores. One could also undertake this iterative procedure until convergence is achieved for the correlation, ρ_1, values using the algorithm outlined in Sect. 7.3.3.2. It is this second approach that we shall now show the results of and accept that convergence has been achieved when the iterative values of ρ_1 are identical to the six decimal place; this is achieved after four iterations.

Table 7.2 Calculating the row scores, $\mathbf{x}_1$ for Table 7.1

Category	Strongly against	Against	Neutral	For	Strongly for
Iteration	$x_1^{(l)}$	$x_2^{(l)}$	$x_3^{(l)}$	$x_4^{(l)}$	$x_5^{(l)}$
$l = 1$	1.237413	0.615025	0.473983	−0.880564	−1.445856
$l = 2$	1.209150	0.636327	0.487018	−0.869946	−1.462550
$l = 3$	1.201807	0.642564	0.489692	−0.867827	−1.466236
$l = 4$	1.199902	0.644221	0.490352	−0.867336	−1.467139

Table 7.3 Calculating the row scores, $\mathbf{y}_1$ for Table 7.1

Category	Never	Rarely	Sometimes	Often	Always
Iteration	$y_1^{(l)}$	$y_2^{(l)}$	$y_3^{(l)}$	$y_4^{(l)}$	$y_5^{(l)}$
$l = 1$	−1.318477	−0.575153	0.436004	0.575820	1.539114
$l = 2$	−1.303532	−0.585411	0.434184	0.577401	1.544244
$l = 3$	−1.299701	−0.588345	0.433770	0.578240	1.545399
$l = 4$	−1.298714	−0.589105	0.433668	0.578487	1.545684

Table 7.4 Calculating the value of ρ_1 for Table 7.1

Iteration	$\rho_1^{(l)}$
$l = 1$	−0.647053
$l = 2$	−0.647200
$l = 3$	−0.647210
$l = 4$	−0.647210

Table 7.2 summarizes the values of $\mathbf{x}_1^{(l)} = \left(x_1^{(l)} \ldots x_5^{(l)}\right)^T$ at the lth iteration, for $l = 1, 2, 3, 4$ while Table 7.3 gives the values of $\mathbf{y}_1^{(l)} = \left(y_1^{(l)} \ldots y_5^{(l)}\right)^T$. Table 7.4 summarizes the values of $\rho_1^{(l)}$ at each of the four iterations.

These results show that the scores that maximize the association between the row and column variables of Table 7.1 are

$$\mathbf{x}_1 = (1.199902,\ 0.644221,\ 0.490352,\ -0.867336,\ -1.467139)^T \tag{7.39}$$

and

$$\mathbf{y}_1 = (-1.298714,\ -0.589105,\ 0.433668,\ 0.578487,\ 1.545684)^T\ , \tag{7.40}$$

respectively, while the correlation between these scores is $\rho_1 = -0.647210$.

Note that if we consider multiplying the row scores by negative one so that

$$\mathbf{x}_1 = (-1.199902,\ -0.644221,\ -0.490352,\ 0.867336,\ 1.467139)^T$$

and leave the column scores unchanged, then the correlation between them is $\rho_1 = 0.647210$. This is important for our discussion of how the singular value decomposition of the matrix $\mathbf{Z}$ can be used for the calculation of these, and higher dimensional, scores. We now describe the results of this decomposition.

7.5.2 K-Dimensional Solution via SVD

Using R, we determine the row and column scores, $\mathbf{x}_1$ and $\mathbf{y}_1$ (and the higher dimensional vectors) by applying the `svd()` function to the matrix $\mathbf{Z}$ given by Eq. (7.38). By doing so, the resulting matrix of row scores is of size $m \times K$ and is defined as

$$\tilde{\mathbf{X}} = (\tilde{\mathbf{x}}_1 \quad \tilde{\mathbf{x}}_2 \quad \ldots \quad \tilde{\mathbf{x}}_K)$$

where $K = \min(m,\ n) - 1 = \min(5,\ 5) - 1 = 4$. Therefore,

$$\tilde{\mathbf{X}} = \begin{pmatrix} -0.5363 & 0.6802 & -0.0329 & -0.2205 \\ -0.2884 & -0.5970 & 0.5142 & -0.3099 \\ -0.2194 & -0.2310 & -0.3132 & 0.7749 \\ 0.3878 & -0.1676 & -0.6419 & -0.4577 \\ 0.6563 & 0.3154 & 0.4737 & 0.2132 \end{pmatrix}.$$

Similarly, the matrix of columns scores $\mathbf{Y}$ of size $n \times K = 5 \times 4$ is

$$\begin{aligned} \tilde{\mathbf{Y}} &= (\tilde{\mathbf{y}}_1 \quad \tilde{\mathbf{y}}_2 \quad \tilde{\mathbf{y}}_3 \quad \tilde{\mathbf{y}}_4) \\ &= \begin{pmatrix} -0.5702 & 0.6797 & -0.1415 & 0.0065 \\ -0.3415 & -0.6916 & 0.1373 & 0.2248 \\ 0.1795 & -0.0648 & -0.2108 & -0.8647 \\ 0.1763 & 0.1236 & -0.8465 & 0.3797 \\ 0.7035 & 0.2008 & 0.4473 & 0.2399 \end{pmatrix}. \end{aligned}$$

We can determine the matrices $\mathbf{X}$ and $\mathbf{Y}$ by pre-multiplying $\tilde{\mathbf{X}}$ and $\tilde{\mathbf{Y}}$ by $\mathbf{D}_r^{-1/2}$ and $\mathbf{Y}$ by $\mathbf{D}_c^{-1/2}$, respectively. Doing so gives

$$\begin{aligned} \mathbf{X} &= \mathbf{D}_r^{-1/2}\tilde{\mathbf{X}} \\ &= \begin{pmatrix} -1.1992 & 1.5210 & -0.0735 & -0.4930 \\ -0.6448 & -1.3349 & 1.1498 & -0.6929 \\ -0.4906 & -0.5165 & -0.7003 & 1.7327 \\ 0.8672 & -0.3749 & -1.4353 & -1.0234 \\ 1.4674 & 0.7053 & 1.0592 & 0.4767 \end{pmatrix} \end{aligned}$$

and

$$\mathbf{Y} = \mathbf{D}_c^{-1/2}\tilde{\mathbf{Y}}$$
$$= \begin{pmatrix} -1.2984 & 1.5477 & 0.3221 & 0.0148 \\ -0.5894 & -1.1936 & 0.2370 & 0.3881 \\ 0.4336 & -0.1564 & -0.5092 & -2.0884 \\ 0.5786 & 0.4055 & -2.7781 & 1.2460 \\ 1.5458 & 0.4411 & 0.9827 & 0.5271 \end{pmatrix}.$$

Note that, with some rounding error introduced, the first column of **X** and **Y** are identical to the vectors $\mathbf{x}_1$ and $\mathbf{y}_1$ that were calculated using the reciprocal averaging procedure. By using the `svd()` function in R, the $K = 4$ singular values of **Z** are $\rho_1 = 0.6472$, $\rho_2 = 0.3290$, $\rho_3 = 0.1730$ and $\rho_4 = 0.0324$ and are all positive and are arranged in descending order. Note that the maximum of these values is equivalent to the maximum correlation ρ_1 we found above. Therefore, the matrix Λ in Eq. (7.28) is

$$\Lambda = \begin{pmatrix} 0.6472 & 0 & 0 & 0 \\ 0 & 0.3290 & 0 & 0 \\ 0 & 0 & 0.1730 & 0 \\ 0 & 0 & 0 & 0.0324 \end{pmatrix}.$$

7.5.3 On Reconstituting the Cell Frequencies

By substituting **X**, **Y** and Λ into Eq. (7.28), we can perfectly reconstruct the joint proportions of Table 7.1; see Eq. (7.37). If we just confined our attention to the one-dimensional solution and consider only $\mathbf{x}_1$ (see Eq. (7.39)), $\mathbf{y}_1$ (see Eq. (7.40)) and $\rho_1 = -0.647210$ calculated using reciprocal averaging then we can get an approximation of Eq. (7.37) by considering Eq. (7.30). By multiplying this approximation of **P** by the sample size, f_t, gives the following one-dimensional approximation of the cell frequencies

$$\mathbf{N} \approx \begin{pmatrix} 10.8463 & 13.7004 & 3.1834 & 1.4320 & -1.1621 \\ 8.3241 & 11.7089 & 3.9321 & 1.9729 & 2.0621 \\ 7.6257 & 11.1574 & 4.1394 & 2.1227 & 2.9549 \\ 1.4632 & 6.2915 & 5.9685 & 3.4443 & 10.8325 \\ -1.2592 & 4.1418 & 6.7766 & 4.0282 & 14.3127 \end{pmatrix}.$$

Generally, reconstituting the cell frequencies using the row scores, column scores, and their correlation, from reciprocal averaging produces quite good results. However, it is clear there are two cell frequencies which are clearly invalid; where the reconstituted frequencies are -1.2592 and -1.1621. One may then consider the exponential form of the model—Eq. (7.31)—which produces the following reconstituted cell frequencies

$$\mathbf{N} \approx \begin{pmatrix} 14.8050 & 14.8530 & 3.4275 & 1.6591 & 1.7463 \\ 9.2803 & 12.0172 & 4.0060 & 2.0428 & 3.0447 \\ 8.1544 & 11.3324 & 4.1828 & 2.1639 & 3.5513 \\ 2.6048 & 6.7532 & 6.1230 & 3.5975 & 13.8118 \\ 1.5734 & 5.3727 & 7.2457 & 4.5033 & 25.1676 \end{pmatrix}.$$

Such a strategy now yields cell frequencies that are all non-negative. There are various issues concerned with reconstituting, or even approximating, the cell frequencies in this manner but we shall not discuss this issue any further.

7.6 Final Remarks

Reciprocal averaging and canonical correlation analysis have long, diverse and interesting histories, and we have only briefly touched upon the key elements of both here. What is apparent from our discussion is that the method of reciprocal averaging involves a specific type of "averaging" of the elements of the centred row and column profiles. We shall examine this issue in more detail in the next chapter and build upon the idea of reciprocal averaging further. However, our discussion in Chap. 8 will be only to provide avenues of possible investigation and do not go any further than that. There is no simulation study, or empirical analysis of these ideas but rather should be read as a "technical essay". Therefore, our discussion of reciprocal averaging will take what Hill (1974) and Nishisato (1980) describe in their descriptions of the method and look at alternative "averaging" strategies of the profiles.

References

Benzécri, J.-P. (1973). *Analyse des Données*. Paris: Dunod.

de Leeuw, J. (1983). On the prehistory of correspondence analysis. *Statistica Neerlandica, 37,* 161–164.

Fisher, R., & Mackenzie, W. A. (1923). Studies in crop variation ii. The manurial response of different potatio varieties. *The Journal of Agricultural Science, 13,* 311–320.

Gifi, A. (1990). *Non-linear multivariate analysis*. Chichester: Wiley.

Gower, J. (1989). Generalized canonical analysis. In R. Coppi & S. Bolasco (Eds.), *Multiway data analysis* (pp. 221–232). Amsterdam: North Holland.

Gower, J. C., Lubbe, S., & Le Roux, N. (2011). *Understanding biplots*. Chichester: Wiley.

Hill, M. (1974). Correspondence analysis: A neglected multivariate technique. *Journal of the Royal Statistical Society (Series, C), 23,* 340–354.

Hirschfeld, H. (1935). A connection between correlation and contingency. *Proceedings of the Cambridge Philosophical Society, 31,* 520–524.

Horst, P. (1935). Measuring complex attitudes. *Journal of Social Psychology, 6,* 369–374.

Lawshe, C., & Harris, D. (1958). The method of reciprocal averages in weighting personnel data. *Educational and Psychological Measurement, 18,* 331–336.

Michailidis, G., & de Leeuw, J. (1998). The Gifi system of descriptive multivariate analysis. *Statistical Science, 13,* 307–336.

Mitzel, H. (1954). A methodological study of reciprocal averages technique applied to an attitude scale. *Journal of Counseling Psychology, 1,* 256–359.

Mosier, C. (1946). Rating of training and experience in public personnel selection. *Educational and Psychological Measurement, 6,* 313–329.

Nishisato, S. (1980). *Analysis of categorical data: Dual scaling and its applications.* Toronto: University of Toronto Press.

Richardson, M., & Kuder, G. F. (1933). Making a rating scale that measures. *Personnel Journal, 12,* 36–40.

Thurstone, L. L. (1928). The measurement of opinion, *Journal of Abnormal and Social Psychology, 22,* 415–430.

Van Rijckevorsel, J. L. A. (1987). *The application of fuzzy coding and horseshoes in multiple correspondence analysis.* Leiden: DSWO Press.

Welsh, A., & Robinson, J. (2005). Fisher and inference for scores. *International Statistical Review, 73,* 131–150.

Chapter 8
Some Generalizations of Reciprocal Averaging

8.1 Introduction

In Chap. 7, we showed how reciprocal averaging and canonical correlation analysis yield identical row and column scores. Our discussions were centred on the role of row and column profiles, and their centred versions, and so it is these profiles that form the foundation of many scoring techniques concerned with the analysis of categorical variables, including correspondence analysis and its analogs. As we have discussed, reciprocal averaging involves applying an "averaging" technique to these profiles, thereby generating measures of central tendency for the categories. There are other ways in which the scores can be determined for the two categorical variables that rely on alternative "averaging" strategies. Therefore, this chapter will discuss amendments to reciprocal averaging that involve alternatives to the standard "averaging" of the centred row and column profiles. In doing so, and for the sake of simplicity, we shall restrict our attention to just the one-dimensional problem. We must point out that our discussions focus only on presenting several different strategies for doing so and do not verify the benefits, or lack thereof, of each of them. So this chapter is to be read as a discussion point and further research will need to be undertaken to better understand the features of these alternative "averaging" approaches and their application.

8.2 Method of Reciprocal Medians (MRM)

Recall that our description of reciprocal averaging in Chap. 7 centred on determining the ith row score, x_{i1} and the jth column score, y_{j1} by "averaging" the centred row and column profiles. That is, we showed that these scores can be obtained iteratively, or equivalently from an eigendecomposition, through the two formulae

S. Nishisato et al., *Modern Quantification Theory*,
Behaviormetrics: Quantitative Approaches to Human Behavior 8,
https://doi.org/10.1007/978-981-16-2470-4_8

$$\rho_1 x_{i1} = \sum_{j=1}^{n} \left(\frac{p_{ij}}{p_{i\bullet}} - p_{\bullet j} \right) y_{j1}$$

and

$$\rho_k y_{j1} = \sum_{i=1}^{m} \left(\frac{p_{ij}}{p_{\bullet j}} - p_{i\bullet} \right) x_{i1} \,.$$

Note that these two equations are just those of Eqs. (7.3) and (7.4) for $k = 1$.

The right-hand side of these two equations is just the *arithmetic mean* of the weighted centred profiles. Such an approach takes into account the magnitude and sign of all elements of the row and column profiles, even those that may be deemed very large or very small. Since "outlying" profiles can play a significant role in the solution that one obtains for x_{ik} and y_{jk}, there may be a need to consider an alternative strategy to determining these scores. One such alternative is to adopt the reciprocal median's approach proposed by Nishisato (1984). Despite the benefits of considering such an approach, it appears to not be widely known. Therefore, we shall briefly outline the method of *reciprocal medians* (or MRM).

Nishisato (1984) argued that, for very large or very small marginal frequencies (or, equivalently, proportions), these categories can dominate the solution obtained when calculating the values of x_{ik} and y_{jk}. Therefore, rather than determining these scores using an arithmetic mean of the elements of the centred profiles, one may instead consider their median. By doing so, Nishisato (1984) demonstrated that x_{ik} and y_{jk} may be determined from the following two formulae:

$$\rho_k x_{ik} = \text{Mdn}_j \left(\left(\frac{p_{ij}}{p_{i\bullet}} - p_{\bullet j} \right) y_{jk} \right) \tag{8.1}$$

and

$$\rho_k y_{jk} = \text{Mdn}_i \left(\left(\frac{p_{ij}}{p_{\bullet j}} - p_{i\bullet} \right) x_{ik} \right) \,. \tag{8.2}$$

Here, "Mdn_j" is the median of the m elements of the weighted centred row profile, while "Mdn_i" is the median of the n elements of the centred column profile. Determining x_{ik} and y_{jk} using Eqs. (8.1) and (8.2) is referred to as the method of *reciprocal medians*. It must also be noted that Nishisato's (1984) description of this approach focuses on a matrix of 0's and 1's but in our exploration of the method, it works equally well for a two-way contingency table where its cell values are joint frequencies.

Suppose we let $\mathbf{x}_k = (x_{1k}, x_{2k}, \ldots, x_{mk})^T$ and $\mathbf{y}_k = (y_{1k}, y_{2k}, \ldots, y_{nk})^T$. Then, the properties that Nishisato (1984) imposes upon the row and column scores are just $\mathbf{r}^T \mathbf{x}_k = 0$ and $\mathbf{x}_k^T \mathbf{D}_r \mathbf{x}_k = 1$ for the row scores—see also Eq. (7.7)—and $\mathbf{c}^T \mathbf{y}_k = 0$ and $\mathbf{y}_k^T \mathbf{D}_c \mathbf{y}_k = 1$ for the column scores—see Eq. (7.8). For the other variations of reciprocal averaging that we describe in this chapter, these properties will be observed for the vector of x_{ik} and y_{jk} values.

Once the row and column scores are determined from Eqs. (8.1) and (8.2), the correlation between them is defined such that

$$\rho_k = \sum_{i=1}^{m}\sum_{j=1}^{n} p_{ij} x_{ik} y_{jk} ,$$

and Nishisato (1984) [p. 145] noted that this value is "substantially smaller" than its corresponding value using the traditional approach to reciprocal averaging; see Eq. (7.11). It was also pointed out that this approach is also prone to convergence problems, mainly due to an oscillating solution at each iteration. Such a feature appears to be present when iteratively calculating the row and column scores. However, our experience has shown that if the convergence criterion is applied to the correlation, and not the values of x_{ik} and y_{jk}, then the problem is not so severe.

8.3 Reciprocal Geometric Averaging (RGA)

As we discussed in the previous section, the method of reciprocal medians takes into account any unusual, or outlying, profiles that are present in the contingency table by determining the median of the set of elements in the ith centred row profile of which the jth element is

$$\left(\frac{p_{ij}}{p_{i\bullet}} - p_{\bullet j}\right) y_{j1} . \tag{8.3}$$

Such a strategy is useful when the arithmetic mean is influenced heavily by these profiles. Rather than considering the median of the centred profile, one may instead consider its natural logarithm. Here we describe three ways in which this can be done and examine the pros and cons of each approach. They are all specific types of *reciprocal geometric averaging* (RGA).

8.3.1 RGA of the First Kind (RGA1)

Suppose we take the natural logarithm of Eq. (8.3). Doing so can only be done for non-negative values of Eq. (8.3) such that

$$\ln\left(\left|\frac{p_{ij}}{p_{i\bullet}} - p_{\bullet j}\right| y_{j1}\right) .$$

The mean value of these terms can be found such that

$$\ln(x_{i1}) = \frac{1}{n}\left(\ln\left(\left|\frac{p_{i1}}{p_{i\bullet}} - p_{\bullet 1}\right| y_{11}\right) + \cdots + \ln\left(\left|\frac{p_{in}}{p_{i\bullet}} - p_{\bullet n}\right| y_{n1}\right)\right)$$

$$= \frac{1}{n}\sum_{j=1}^{n} \ln\left(\left|\frac{p_{ij}}{p_{i\bullet}} - p_{\bullet j}\right| y_{j1}\right)$$

where $\ln(x_{i1})$ is the natural logarithm of x_{i1}. This expression may then be rewritten so that

$$x_{i1} = \left(\prod_{j=1}^{n} \left|\frac{p_{ij}}{p_{i\bullet}} - p_{\bullet j}\right| y_{j1}\right)^{1/n}$$

and is the geometric mean of the n elements of the ith row profiles. Taking into consideration the correlation term, ρ_1, we have

$$\rho_1 x_{i1} = \left(\prod_{j=1}^{n} \left|\frac{p_{ij}}{p_{i\bullet}} - p_{\bullet j}\right| y_{j1}\right)^{1/n} . \qquad (8.4)$$

Similarly, we can also show that

$$\rho_1 y_{j1} = \left(\prod_{i=1}^{m} \left|\frac{p_{ij}}{p_{\bullet j}} - p_{i\bullet}\right| x_{i1}\right)^{1/m} . \qquad (8.5)$$

Equations (8.4) and (8.5) form the method of *reciprocal geometric averaging*. We consider alternative versions of this analysis and so we refer to this variant more specifically as the method of *reciprocal geometric averaging of the first kind* (RGA1).

8.3.2 RGA of the Second Kind (RGA2)

A feature of RGA1 is that one must consider the absolute value of the elements of the centred profiles to avoid any problems arising from calculating their natural logarithm when any of their elements are negative. One way to resolve this problem is to consider the following amendment. Rather than taking the natural logarithm of the elements of the centred profiles suppose we instead consider the logarithm of the uncentred profile elements. That is, for the jth element of the ith (uncentred) row profile, we calculate

$$\ln\left[\left(\frac{p_{ij}}{p_{i\bullet}}\right) y_{j1}\right] .$$

When there is complete independence between the row and column variables, this transformation can be centred by considering

$$\ln\left[\left(\frac{p_{ij}}{p_{i\bullet}}\right)y_{jk}\right] - \ln\left[p_{\bullet j}\right] = \ln\left[\left(\frac{p_{ij}}{p_{i\bullet}p_{\bullet j}}\right)y_{j1}\right].$$

Here $p_{ij}/\left(p_{i\bullet}p_{\bullet j}\right)$ is the Pearson ratio of the $(i,\ j)$th cell of the contingency table; see, for example, Goodman (1996), Beh (2004) and Beh and Lombardo (2014). Following the same argument we used to obtain Eqs. (8.4) and (8.5), we get the centred version of *reciprocal geometric averaging*

$$\rho_1 x_{i1} = \left[\prod_{j=1}^{n}\left(\frac{p_{ij}}{p_{i\bullet}p_{\bullet j}}\right)y_{j1}\right]^{1/n} \quad (8.6)$$

and

$$\rho_1 y_{j1} = \left[\prod_{i=1}^{m}\left(\frac{p_{ij}}{p_{i\bullet}p_{\bullet j}}\right)x_{i1}\right]^{1/m}. \quad (8.7)$$

Therefore, the right-hand side of Eqs. (8.6) and (8.7) is the geometric mean of the Pearson ratio. This variant of reciprocal averaging is referred to as the method of *reciprocal geometric averaging of the second kind* (RGA2).

8.3.3 RGA of the Third Kind (RGA3)

A feature of RGA1 and RGA2 is that when calculating their row and columns scores, the uncentred versions of x_{i1} and y_{j1} are always positive. A way around this is to remove the scores from within the natural logarithm term and consider instead

$$\begin{aligned} y_{j1}\ln\left(\frac{p_{ij}}{p_{i\bullet}}\right) - y_{j1}\ln\left(p_{\bullet j}\right) &= y_{j1}\ln\left(\frac{p_{ij}}{p_{i\bullet}p_{\bullet j}}\right) \\ &= \ln\left(\frac{p_{ij}}{p_{i\bullet}p_{\bullet j}}\right)^{y_{j1}}. \end{aligned}$$

Doing so produces the reciprocal averaging formula

$$\rho_1 x_{i1} = \prod_{j=1}^{n}\left(\frac{p_{ij}}{p_{i\bullet}p_{\bullet j}}\right)^{y_{j1}}. \quad (8.8)$$

By the same argument,

$$\rho_1 y_{j1} = \prod_{i=1}^{m}\left(\frac{p_{ij}}{p_{i\bullet}p_{\bullet j}}\right)^{x_{i1}}. \quad (8.9)$$

The advantage of this strategy is that when there is complete independence between the variables of a contingency table, then this term will be zero. When this is the case, $x_{i1} = 1$, for $i = 1, 2, \ldots, m$ and $y_{j1} = 1$, for $j = 1, 2, \ldots, n$ are the trivial solutions to the row and column scores so that $\rho_1 = 1$.

While Eqs. (8.8) and (8.9) do not, strictly speaking, involve the calculation of the geometric mean of the Pearson ratio, they are quite similar in form and, together, are the formulae used to calculate x_{i1} and y_{j1} using the method of *reciprocal geometric averaging of the third kind* (RGA3).

8.4 Reciprocal Harmonic Averaging (RHA)

Suppose we consider again the jth element of the ith centred row profiles

$$\left(\frac{p_{ij}}{p_{i\bullet}} - p_{\bullet j}\right) y_{j1} .$$

Recall that the traditional approach to reciprocal averaging involves the (unweighted) sum of these values. Suppose we now consider a weighted version so that

$$\begin{aligned}\rho_1 x_{i1} &= \frac{w_1\left(\frac{p_{i1}}{p_{i\bullet}} - p_{\bullet 1}\right) y_{11} + \cdots + \left(\frac{p_{in}}{p_{i\bullet}} - p_{\bullet n}\right) y_{n1}}{w_1 + \cdots + w_n} \\ &= \frac{\sum_{j=1}^{n} w_j \left(\frac{p_{ij}}{p_{i\bullet}} - p_{\bullet j}\right) y_{j1}}{\sum_{j=1}^{n} w_j}\end{aligned}$$

for some unconstrained weight w_j, $j = 1, 2, \ldots, n$. Let

$$d_{j|i} = w_j \left(\frac{p_{ij}}{p_{i\bullet}} - p_{\bullet j}\right) y_{j1} .$$

Then, since the w_j values are unconstrained, there will always be a value such that $d_{j|i}$ is constant for all $j = 1, 2 \ldots, n$. Therefore, by assuming $d_{j|i}$ is constant for all values of j, so that

$$d_{\bullet|i} = d_{j|i}$$

then

$$
\begin{aligned}
\rho_1 x_{i1} &= \frac{\sum_{j=1}^{n} d_{\bullet|i}}{\sum_{j=1}^{n} d_{\bullet|i} / \left(\left(\frac{p_{ij}}{p_{i\bullet}} - p_{\bullet j} \right) y_{j1} \right)} \\
&= \frac{n d_{\bullet|i}}{d_{\bullet|i} \sum_{j=1}^{n} \frac{1}{\left(\frac{p_{ij}}{p_{i\bullet}} - p_{\bullet j} \right) y_{jk}}}
\end{aligned}
$$

which simplifies to

$$
\rho_1 x_{i1} = n \left(\sum_{j=1}^{n} \frac{1}{\left(\frac{p_{ij}}{p_{i\bullet}} - p_{\bullet j} \right) y_{j1}} \right)^{-1} . \tag{8.10}
$$

One can see here that the right-hand side of Eq. (8.10) is the harmonic mean of the elements of the ith centred row profile. By applying a similar strategy to the elements of the centred column profiles, we obtain

$$
\rho_1 y_{j1} = m \left(\sum_{i=1}^{m} \frac{1}{\left(\frac{p_{ij}}{p_{\bullet j}} - p_{i\bullet} \right) x_{i1}} \right)^{-1} \tag{8.11}
$$

where the right-hand side is the harmonic mean of the elements of the jth centred column profiles. Together, Eqs. (8.10) and (8.11) define the *reciprocal harmonic averaging* of the row and column variables of a two-way contingency table.

8.5 Final Remarks

The scaling of the row and column categories of a contingency table can be undertaken in a number of ways and in Chap. 7 we described how this can be achieved using reciprocal averaging. We also showed how reciprocal averaging can be linked to the eigendecomposition of a matrix involving the Pearson standardized residual where the $(i,\ j)$th element is

$$
Z_{ij} = \frac{p_{ij} - p_{i\bullet} p_{\bullet j}}{\sqrt{p_{i\bullet} p_{\bullet j}}} .
$$

A feature of reciprocal averaging is that it involves the arithmetic averaging of the elements of the centred row and column profiles. Therefore, the aim of this chapter has to present a variety of ways in which reciprocal averaging can be modified to incorporate other "averaging" techniques. We freely admit that this chapter only presents some strategies and does not go any further than that. We have not presented any demonstration of the application of these methods, nor have we compared the scaling (and correlation) obtained from using them. Thus, there is plenty of scope to examine the pros and cons of each of these and to seek ways in which they may present a better way of scaling than the traditional approach to reciprocal averaging.

Another aspect of our discussion here is that we have also made no attempt to link these reciprocal averaging techniques to the eigendecomposition of a matrix or to any form of canonical correlation analysis. So there is plenty of room left to discuss these linkages as well.

This chapter marks the end of our discussion of scaling issues for a two-way contingency table. We now turn our attention to the benefits of visualizing the association between categorical variables, and there are many ways of doing so that use row and column scores (found via reciprocal averaging, eigendecomposition, singular value decomposition, or by other means) as the foundation for these visual summaries. In Part I of this book, numerous discussions have been made about the pros and cons of a variety of plotting strategies including the utility of joint displays. In the chapters that are to follow, we shall describe some of the history, development, and application of the biplot for visually summarizing relationships/associations that exist between numerical and categorical variables.

References

Beh, E. J. (2004). Simple correspondence analysis: A bibliographic review. *International Statistical Review, 72*, 257–284.

Beh, E. J., & Lombardo, R. (2014). *Correspondence analysis, practice, methods and new strategies*. Chichester: Wiley.

Goodman, L. A. (1996). A single general method for the analysis of cross-classified data : Reconciliation and synthesis of some methods of pearson, yule, and fisher, and also some methods of correspondence analysis and association analysis. *Journal of the American Statistical Association, 91*, 408–428.

Nishisato, S. (1984). Dual scaling by reciprocal medians. *Atti della XXXII Riunione Scientifica della Societá Italiana di Statistica,* 141–147.

Chapter 9
History of the Biplot

9.1 Introduction

Visualization techniques represent one of the main pillars in the field of exploratory data analysis. A graphical description of data is often a more preferred option than a numerical one as it is more intuitive and immediate. Boxplots, histograms, and pie charts are familiar forms of data visualization which require only a rudimentary statistical understanding to construct and interpret. However, they are applicable only to univariate data. For multivariate data sets, a biplot can be presented in a manner which can be readily understood by non-statistically minded individuals. Ancient and primitive art, like African sculpture and Picasso's invention of Cubism, offer examples as to how *natural forms* or data structures/types/configurations can be reduced to purely geometrical equivalents (Loach 1998). While addressing a group of architecture students at Columbia University, Le Corbusier is recorded as saying in 1961

> ...I prefer drawing to talking. Drawing is faster, and allows less room for lies.

This is especially so with the advances constantly being made in computer technology and graphical displays. Like biplots, drawing helps to *get a feel* for the data rather than going straight to model building and testing hypotheses which is often the case in the application of a range of statistical tools and techniques.

When analysing multivariate data, a versitile and popular means of data visualization is the *biplot*. As Gower et al. (2015)[p. 1 Abstract] states

> Biplots provide visualizations of two things, usually, but not necessarily, in two dimensions

adding that (page 1)

> . . . the *bi-* of biplots refers to the two modes and not the usual two dimensions used for display.

S. Nishisato et al., *Modern Quantification Theory*,
Behaviormetrics: Quantitative Approaches to Human Behavior 8,
https://doi.org/10.1007/978-981-16-2470-4_9

Here, a *mode* is the variable or sample of individuals being analysed. Therefore, a biplot is a plot of two *things* and can be seen as a direct generalization of the familiar *scatter plot* of two variables to the visualization of many variables in a geometrical space of reduced dimension. In contrast to the scatter plot, the axes are not perpendicular, since biplots consider the projection of a n-dimensional representation over a surface with a minimum loss of information.

Interestingly, Gower et al. (2011) describes that the basic idea of a biplot can be traced back to Ptolemy's map which provides a two-dimensional visualization of the known world during the Hellenistic period in the second century. Indeed, Ptolemy's map of the old world was deemed a realistic representation at the time and considered two important *things* (1) points that depict the location of cities and (2) lines that represent latitude and longitude. The use of the term *map* is not just confined to Ptolemy but has also been adopted in the analysis of multivariate data. For example, Greenacre (1984) refers to a biplot as a map of the data, providing no geographical connotation but instead describing such a map from a geometrical perspective.

The revolutionary idea of linking coordinates that are algebraically derived to visualize points in a low-dimensional space can be traced back to Descartes (1637). More recent history tells us that biplots were first introduced by Gabriel (1971) for continuous variables and for portraying the relationship between individuals and variables described in principal component analysis (PCA) (Pearson 1901; Hotelling 1933). Gower and Hand (1996) and, later, Gower et al. (2011) wrote interesting monographs on biplots. Yan and Kang (2003) also described various approaches that can be used to visualize and interpret a biplot. Greenacre (2010a) wrote a practical user-oriented guide to biplots, focusing on their application to a variety of techniques including principal component analysis (PCA), multidimensional scaling (MDS), log-ratio analysis (LRA, also known as spectral mapping), and discriminant analysis (DA). Greenacre (2010a) also described the biplot for some variants of correspondence analysis, such as simple correspondence analysis (CA), multiple correspondence analysis (MCA), and canonical correspondence analysis (CCA). Gower et al. (2011) gives an excellent description of the biplot which helps to enhance its popularity as a useful tool for the visualization of multivariate data.

Biplots are not primarily a method of analysis, but are instead a convenient way of portraying numerical or categorical information observed on a great number of individuals, which may or may not derive from previous analyses. They have been described and applied in many ways and in very different scientific areas such as medicine (Gabriel and Odoroff 1990), genetics (Wouters et al. 2003), agriculture (Yan et al. 2000), library science (Veiga de Cabo and Martín-Rodero 2011), economics and business (Galindo et al. 2011; Galindo 1986), tourism (Pan et al. 2008), political science (Alcántara and Rivas 2007), and bibliometrics (Arias Díaz-Faes et al. 2011; Torres-Salinas et al. 2013). There also exists a wide variety of different types of biplots including the nonlinear biplot (Gower and Harding 1988; Groenen et al. 2015), contribution biplots (Greenacre 2013), interactive biplots for use with `XLISP-STAT` (Udina 2005), and biplots for the analysis of compositional data (Aitchison and Greenacre 2002). Furthermore, biplots have been used in a variety of difference statistical contexts and are briefly sketched out here.

Calibrated Biplots

Calibrated biplots, originally proposed by Gabriel and Odoroff (1990) and presented in a correspondence analysis framework by Carlier and Kroonenberg (1996) and Lombardo et al. (1996), have been largely discussed by Gower et al. (2011). For this type of biplot, the approximated values of a data table can be directly read from the graph resulting in a *calibration* of a biplot axis. For further details see Gower et al. (2011), Gower et al. (2015), Gower et al. (2016), Greenacre (1993a, b, 2010a), and Gower and Hand (1996).

Biplots for Log-Ratio Analysis

Biplots have also been used for the visualization of strictly positive data, log-transformed, and double-centred data, using singular value decomposition (SVD). They have also been used in log-ratio analysis using (again) SVD; see Aitchison and Greenacre (2002) and Greenacre (2010b). The method is very useful for the analysis of compositional data (Aitchison 1983, 1986). There have been two forms of this biplot proposed in the log-ratio analysis literature; an unweighted form (see Aitchison 1983 and Aitchison and Greenacre 2002) and a weighted form, with the latter also being referred to as *spectral mapping* (Greenacre and Lewi 2009; Lewi 1976).

Biplots for Discriminant Analysis

A linear discriminant analysis (LDA) and multivariate analysis of variance (MANOVA) can be decomposed using a SVD of the matrix of group centroids, weighted by their respective group sizes, in a space structured by the Mahalanobis metric—see, for example, Mardia et al. (1979). Therefore, it is possible to portray the variable relationship through biplots where the principal axes are often referred to as *canonical axes*; the terminology *canonical variate analysis* is a common synonym used for this type of analysis. The idea of a MANOVA biplot originates with the work of Gabriel (1972).

Biplots and Regression

Interestingly, a biplot can be used for portraying variables in a linear regression model where the coordinates of the axes/vectors can be shown to be regression coefficients of the variables and the coordinates of points as the standardized independent variables; see Greenacre (2010a) [p. 213–217].

Biplots and RC(M) Association Model

De Rooij and Heiser (2005) presented the distance-association models and discussed two types of graphical representations that jointly display the rows and columns of a contingency table when studying the association between two categorical variables. The two types of plots they described for this analysis are the *type I plots* which provides a visualization where the relationship between the two sets is described by a distance rule, and the *type II plot* that visualizes this relationship using an inner product rule. The construction of the type II plot is consistent with the construction of a biplot where one set of points is visualized using vectors and the second set of points is represented by points projected onto these vectors.

Generally, the RC(M) association model described by Goodman (1979, 1985) can be used to produce a graphical representation of the association between the variables based on inner products, using a SVD of the matrix containing a basic set of log-odds ratios (Agresti 2002, p.18).

9.2 Biplot Construction

In line with Pearson approach to PCA (Pearson 1901), we focus on approximating a (standardized) data matrix $\mathbf{Z}$ of size $N \times p$ in a low-dimensional space. We also view the biplot as a multivariate analogue of an ordinary scatter plot where the rows of $\mathbf{Z}$ are represented in a biplot as *points* belonging to a p-space (column-space) and the variables as *axes* of a N-space (row-space). However, whether the biplot is viewed from Gabriel's definition (for the visualization of numerical data) or from Gower's perspective (which includes the visualization of categorical variables), it is important to note that it is based on an inner product between the *points* and the *axes* which allows one to reproduce $\mathbf{Z}$. Simple applications of the biplot can be considered for different variants of correspondence analysis (Beh and Lombardo 2014, 2020); we shall discuss these further in Chap. 10 where the ordered and the three-way forms of CA will also be discussed.

When determining a low-dimensional approximation of a numerical or categorical data matrix, SVD (Eckart and Young 1936) provides a very convenient approach for data reduction; see Sect. 7.3.5. Here, we refer to the *points* using the term *principal coordinates* while the *axes* are projections from the origin to points defined by their *standard coordinates*. Both these terms are widely used in the framework of both PCA and CA.

When analysing numerical data, a biplot is constructed by applying a SVD to a transformed version of a data matrix $\mathbf{Z}$ (usually centred and standardized) to obtain a low-rank approximation whose N rows are the samples (or observations made on individuals, subjects or objects) and whose p columns are the numerical variables.

Assuming that the variables observed on the N individuals are of different scale, or measured using different units, a PCA can be performed on this data. To do so, we first need to define the original data matrix $\hat{\mathbf{Z}}$ and then the matrix $\mathbf{Z}$ where the variables are standardized. Since the rows of the $\hat{\mathbf{Z}}$ reflect each of the N individuals in the sample, we assign uniform weights to them. Each individual is assigned a weight $1/N$. Since there will be variability in each of the p numerical variables, the weight (also referred to as metric) given to the jth column is its standard deviation, σ_j, and its inverse is the (j, j)th element of the diagonal matrix $\mathbf{D}_\sigma = \text{diag}\left(1/\sigma_j\right)$, while $\overline{\hat{Z}}_j$ is the mean of the observations for the jth variable. Therefore, $\hat{\mathbf{Z}}$ can be standardized such that, after standarding each of the numerical variables, the (i, j)th element is defined by

$$Z_{ij} = \frac{\hat{Z}_{ij} - \overline{\hat{Z}}_j}{N\sigma_j}$$

so that

$$\mathbf{Z} = \frac{1}{N}\hat{\mathbf{Z}}\mathbf{D}_\sigma \, .$$

The three most important factorizations of **Z** have been based on the SVD of the matrix. The use of SVD leads to some interesting and convenient properties and, for the three decompositions, are

- the row isometric factorization that takes on the form

$$\begin{aligned}\mathbf{Z} &= \mathbf{R}\left(\mathbf{C}^s\right)^T \\ &= \tilde{\mathbf{X}}\Lambda\tilde{\mathbf{Y}} \, .\end{aligned}$$

The resulting *row isometric biplot* or simply *row biplot* is constructed where $\mathbf{R} = \tilde{\mathbf{X}}\Lambda$ is the column matrix of row principal coordinates and $\mathbf{C}^s = \tilde{\mathbf{Y}}$ is the column matrix consisting of the column standard coordinates. For this factorization, $\tilde{\mathbf{X}}$ and $\tilde{\mathbf{Y}}$ have the property $\tilde{\mathbf{X}}^T\tilde{\mathbf{X}} = \mathbf{I}$ and $\tilde{\mathbf{Y}}^T\tilde{\mathbf{Y}} = \mathbf{I}$, respectively. The $(i, \ j)$th element of this factorization is therefore of the form

$$\begin{aligned}Z_{ij} &= \sum_{k=1}^{K}(\rho_k\tilde{x}_{ik})\tilde{y}_{jk} \\ &= \sum_{k=1}^{K} r_{ik}c^s_{jk} \, .\end{aligned}$$

By using this factorization, the joint display projects the column standard coordinates from the origin on the same space as the row points displayed using their principal coordinates is also referred to as a *JK biplot* or a *form biplot*; see Torres-Salinas et al. (2013) and Aitchison and Greenacre (2002), respectively.

- the column isometric factorization takes on the form

$$\begin{aligned}\mathbf{Z} &= \mathbf{R}^s\mathbf{C}^T \\ &= \tilde{\mathbf{X}}\Lambda\tilde{\mathbf{Y}} \, .\end{aligned}$$

The resulting *column isometric biplot* or simply *column biplot* is constructed where $\mathbf{R}^s = \tilde{\mathbf{X}}$ is the column matrix of row standard coordinates and $\mathbf{C} = \tilde{\mathbf{Y}}\Lambda$ is the column matrix consisting of column principal coordinates. Like we saw for the row isometric factorization, $\tilde{\mathbf{X}}$ and $\tilde{\mathbf{Y}}$ have the property $\tilde{\mathbf{X}}^T\tilde{\mathbf{X}} = \mathbf{I}$ and $\tilde{\mathbf{Y}}^T\tilde{\mathbf{Y}} = \mathbf{I}$, respectively. Also, from this factorization, the $(i, \ j)$th element is

$$
\begin{aligned}
Z_{ij} &= \sum_{k=1}^{K} \tilde{x}_{ik}(\rho_k \tilde{y}_{jk}) \\
&= \sum_{k=1}^{K} r_{ik}^{s} c_{jk}
\end{aligned}
$$

and the joint representation of the coordinates is also referred to as a *GH biplot*; see Torres-Salinas et al. (2013).

- the third factorization that can be used to generate a biplot is the symmetric factorization and leads to the *symmetric biplot*. For this biplot, the matrix $\mathbf{Z}$ is factorized so that

$$
\begin{aligned}
\mathbf{Z} &= \hat{\mathbf{R}}\hat{\mathbf{C}}^T \\
&= \tilde{\mathbf{X}}\Lambda\tilde{\mathbf{Y}}
\end{aligned}
$$

where $\hat{\mathbf{R}} = \tilde{\mathbf{X}}\Lambda^{1/2}$ and $\hat{\mathbf{C}} = \tilde{\mathbf{Y}}\Lambda^{1/2}$, observing that $\tilde{\mathbf{X}}$ and $\tilde{\mathbf{Y}}$ have the property $\tilde{\mathbf{X}}^T\tilde{\mathbf{X}} = \mathbf{I}$ and $\tilde{\mathbf{Y}}^T\tilde{\mathbf{Y}} = \mathbf{I}$, respectively. The (i, j)th of $\mathbf{Z}$ can be expressed using sigma notation such that

$$
\begin{aligned}
Z_{ij} &= \sum_{k=1}^{K} (\sqrt{\rho_k}\tilde{x}_{ik})(\sqrt{\rho_k}\tilde{y}_{jk}) \\
&= \sum_{k=1}^{K} \hat{r}_{ik}\hat{c}_{jk} .
\end{aligned}
$$

The resulting biplot from this factorization is also fairly referred to as a SQRT biplot; see Torres-Salinas et al. (2013).

The construction of these three biplots can be expressed in a more general fashion by defining the matrix of row and column coordinates as

$$
\mathbf{R} = \tilde{\mathbf{X}}\Lambda^{\gamma}
$$

and

$$
\mathbf{C} = \tilde{\mathbf{Y}}\Lambda^{1-\gamma}
$$

where $\gamma = 1$, $\gamma = 0$ and $\gamma = 1/2$ produce the row isometric, column isometric and symmetric biplots, respectively. Of course, different values of γ may be considered and produce biplots with different within-set interpretations of the row and column points. The most common choices of γ—being $\gamma = 1$ and $\gamma = 0$—are the most convenient and produce a plot with a simple interpertation. Part of the reason for this is because the distance measures between the principal coordinates in the resulting biplots are Euclidean.

Once a SVD is applied to **Z** (as we have done to produce the three biplots just described) a least-squares matrix approximation of **Z** can be obtained for any given rank K of **Z**. Eckart and Young (1936) provides an excellent discussion for how to obtain such an approximation and describes that it can be achieved by using the first K singular values of **Z** that are summarized along the diagonal of Λ, and the corresponding K left and right singular vectors contained in the first K columns of $\tilde{\mathbf{X}}$ and $\tilde{\mathbf{Y}}$, respectively. The larger K is, the better the approximation will be with $K = \min(m, n) - 1$ yielding a perfect reconstruction of **Z**.

There are some important features that one should take into account when analyzing the association between categorical variables using a biplot representation. One such feature is the interpretation of distances between principal coordinates, and the nearest distance of a principal coordinate from the projection made of the standard coordinate from the origin. Another important feature that should always be considered is the quality of the graphical representation constructed for visually summarizing the association. This can be done by quantifying the cumulative sum-of-squares of the singular values of **Z** in a two-, or three-, dimensional space, and assessing the percentage contribution this quantity has to the total inertia of the numerical table; for symmetrically associated categorical variables, the total inertia is quantified by dividing the Pearson chi-squared statistic by the sample size.

It is worth noting that, irrespective of the type of variables being studied, a row (or column) isometric biplot is constructed so that all points are projected in the same row (or column) space. For example, suppose we are studying numerical variables that are summarized in a data structure where the rows reflect the individuals in the sample and the columns are the variables being assessed. Then, a row isometric biplot can be constructed so that individuals (rows) are plotted using principal coordinates and the variables (columns) are plotted in the same space using standard coordinates when $\gamma = 1$ yielding the matrix of row principal and column standard coordinates

$$\mathbf{R} = \tilde{\mathbf{X}}\Lambda$$
$$\mathbf{C}^s = \tilde{\mathbf{Y}},$$

respectively. Observe that, when the factorization is of full rank, the individual (row) principal coordinates can be written as

$$\begin{aligned}\mathbf{R} &= \tilde{\mathbf{X}}(\tilde{\mathbf{X}}^T\mathbf{Z}\tilde{\mathbf{Y}}) \\ &= \mathbf{Z}\tilde{\mathbf{Y}}\end{aligned}$$

which shows that these coordinates also belong to the same space as the column variables which are depicted as points defined by their matrix of standard coordinates, $\mathbf{C}^s$.

We now turn our attention to describing more of the features of the biplot when analysing data characterized by numerical variables. In Chap. 10, we extend this discussion to the cross-classification of two, or more, categorical variables that form a contingency table.

9.3 Biplot for Principal Component Analysis

Principal component analysis (PCA) is applied to data that are from numerical variables that are often standardized. This is to ensure a comparison of results can be made when the variables are of a different scale, or measured using different units. Although, when the variables are all of the same scale, or unit, standardization does not need to be undertaken.

Suppose we have a sample of individuals of size N and observations on each individual in this sample are made based on p numerical variables so that we then study the standardized data matrix $\mathbf{Z}$ that summarises this data.

To illustrate this process, and to demonstrate the construction and interpretation of a biplot in the context of PCA, consider the data of Ceyhun et al. (2020). These data stem from a study undertaken to analyze different economic policy measures adopted by 166 countries as a response to the COVID-19 pandemic; here, we shall focus only on the data from 15 selected countries and include the number of positive cases to coronavirus and the number of deaths recorded as of 14 April 2020. These cases and deaths due to COVID-19 are measured relative to the population density of each country. Therefore, Table 9.1 also includes the two variables labelled *casesP* (number of positive cases divided by population density) and *deathsP* (number of deaths divided by the population density). Of these 15 countries, six have a female leader; those being Denmark, Finland, Germany, Iceland, New Zealand, and Norway. Note that of these six countries, all (except Germany) are small with a relatively low population density; Germany is the largest by area, population, and GDP. Therefore,

Table 9.1 Coronavirus and economy

Country	*FiscGDP*	*RateCut*	*MacrFinGDP*	*CasesP*	*DeathP*
Denmark	5.3	−20.0	0.0	22.43	2.59
Finland	1.0	0.0	7.3	110.50	5.22
Germany	4.8	0.0	12.5	223.55	19.37
Iceland	7.8	43.2	1.0	144.67	3.00
New Zealand	5.4	75.0	8.9	25.22	0.67
Norway	2.2	83.3	0.0	463.13	11.00
Spain	1.0	0.0	7.3	1040.12	232.47
United States	10.5	100.0	0.0	18295.86	60.72
Italy	1.7	0.0	7.3	525.71	115.09
Netherlands	2.3	0.0	7.3	50.22	7.25
Japan	4.9	0.0	0.3	27.13	0.69
Australia	9.7	0.7	4.7	763.33	23.67
Canada	6.0	57.1	2.6	5406.50	396.75
China	1.2	0.0	14.1	13.11	30.27
Brazil	3.5	28.5	3.2	886.64	98.48

our aim is to explore the relationships between the selected variables with respect to those countries with a female leader and those countries that have a male leader. Avivah Wittenberg-Cox said, writing in Forbes magazine on 13 April 2020, that

> ... This pandemic is revealing that women have what it takes when the heat rises in our Houses of State. Many will say these are small countries, or islands, or other exceptions. But Germany is large and leading, and the U.K. is an island with very different outcomes. These leaders are gifting us an attractive alternative way of wielding power.

So, we are very much interested in understanding how much truth there is to this statement. To do so there are two *things* (quoting the terminology adopted by Gower et al. 2011), identified in the study of Ceyhun et al. (2020). The first is the *Country* under examination while the second *thing* is the set of *Economic-Pandemic Measures* that were taken. These measures are a collection of five variables that were observed from the economic policy package database of the International Monetary Fund (IMF; Ceyhun, Gokce and Abdullah 2020). The data are summarized in Table 9.1 and include three policy variables that are classified as lying within the set of fiscal policies or the set of monetary policies. The fiscal policy package includes all of the adopted fiscal measures and is coded as a percentage of the gross national product (GDP); this variable is labelled *FiscGDP* in Table 9.1. There are two monetary policy categories being examined in Table 9.1; they include the interest rate cuts by the monetary policy authority, measured as a percentage of the ongoing interest rate on 1 February 2020 (labelled *RateCut*), and the size of the macro-financial package, as a percentage of GDP (labelled *MacrFinGDP*).

The success, or otherwise, or a countries response to the COVID-19 pandemic is assessed by examining these variables in context to the number of cases and deaths recorded in each country. We concede, however, that limiting our attention to the variables in Table 9.1 provides only a partial interpretion of the complex realities faced by leaders' and the varied ways in which they communicate their message to their citizens. The variables we consider do not reflect such the breadth of issues faced by each country, but merely provide a glimpse into their complexities.

Figure 9.1 visually summarizes the relationship of each of the five variables in Table 9.1 with the 15 countries studied using a row isometric biplot. It does so by depicting the countries using point whose position is defined by their principal coordinate while the variables are depicted as projections from the origin to their position defined by their standard coordinate. To distinguish those countries whose leader is a female, the label of their country is in red, while those countries given a blue label in the biplot have male leaders. To interpret how the countries and variables are related from this biplot, we observe how far a *Country* point is located from the projection of a variable; the closer a principal coordinate is to a projection, the greater the correlation between that country and variable. The PCA is carried out using the R package `FactoMiner` (Husson et al. 2010).

The biplot of Fig. 9.1 shows that those six countries who have a female leader have been more generally consistent in their fiscal and financial response to COVID-19 than many of the remaining countries studied. Countries with male leaders that have also displayed such features include Australia, Japan, and The Netherlands.

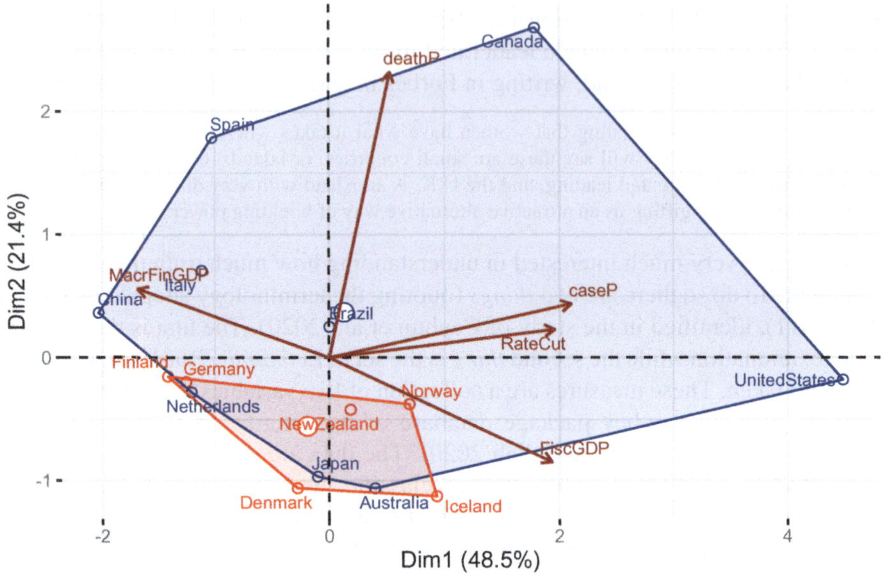

Fig. 9.1 Coronavirus and economical-financial measures

We do note that Germany and Finland are positioned very closely to each other and are very different compared with Iceland and Norway. This difference might be due to the different strategies these countries have taken. For example, Germany and Finland have engaged in dominant actions that are of a macro-financial nature while the policies of Iceland and Norway have been aimed at reducing taxes and interest rates. Such conclusions are apparent when noting the relative closeness of Iceland and Norway, say to the projection of *FiscGDP* and, to a lesser extent, *RateCut*, while the location of Finland and Germany is close to the projection of *MacrFinGDP*. The strategies taken by Australia, New Zealand, and Japan have involved a mix of reducing taxes and macro-financial decisions that impact upon their economic policies.

There are further additional features that the row isometric biplot of Fig. 9.1 provides. The response of the United States (whose leader is male), like Iceland and Norway (whose leaders are female), is very much dominated by cutting interest rates and taxes, while China and Italy (whose leader are males) have taken action that impacts upon their macro-financial policies. Furthermore, Fig. 9.1 shows that Canada and Spain have the highest number of deaths (relative to population density) and, despite actions that appear dominated to impact upon their macro-financial policies, have still struggled financially during the pandemic. Denmark, Japan, Australia, and Iceland, on the other hand, have far fewer deaths (relative to population density).

An important feature of the biplot is the quality of the relationships that it displays in its two dimensions. As it is shown in Fig. 9.1, the first axis accounts for 48.5% of the relationship between the countries and the five variables while the second axis accounts for 21.4% of this relationship. Therefore, Fig. 9.1 visually describes about 70% of the relationship between the two *things*. This is considered a good quality visual display and can be further improved if a third dimension were to be added to the biplot.

It is important to keep in mind that the data summarized in Table 9.1 is valid as of 14 April 2020, and at the time of writing, the impact of COVID-19 globally has impacted each of these countries in different ways; some for the better (such as New Zealand) and some for the worse (including the United States). Therefore, while these data were collected at a relatively early stage of the global pandemic, it may be too premature to conclude that a female leader of a country generally has better (or perhaps more consistently) managed the pandemic than their male counterparts. These results provide a glimpse into how these leaders compare three months into the global response to COVID-19 in terms of the timely economic and financial decisions that needed to be made.

9.4 Final Remarks

In this chapter, we have attempted to provide an overview of the types of biplots that one may construct, the advantages of doing so, and demonstrated their application by analysing numerical variables. The use of biplots has certainly been a topic of increasing discussion and application in the variety of exploratory data analysis techniques that now exist. Of particular relevance is the construction of a biplot for depicting the association between categorical variables using correspondence analysis, at whose heart lies the various scoring strategies discussed in this book. The next chapter will discuss the utility of biplots for the analysis of categorical data and our focus will be to study the association between the variables of two-way and three-way contingency tables.

References

Agresti, A. (2002). *Categorical Data Analysis* (2nd ed.). New York: Wiley.

Aitchison, J. (1983). Principal component analysis of compositional data. *Biometrika, 70,* 57–65.

Aitchison, J. (1986). *The Statistical Analysis of Compositional Data.* London: Chapman & Hall.

Aitchison, J., & Greenacre, M. (2002). Biplots of compositional data. *Journal of the Royal Statistical Society: Series C (Applied Statistics), 51,* 375–392.

Alcántara, M., & Rivas, C. (2007). Las dimensiones de la polarizacion partidista en america latina. *Política y Gobierno, 14,* 349–390.

Arias Díaz-Faes, A., Benito-García, N., Martín-Rodero, H., & Vicente-Villardón, J. (2011). Propuesta de aplicabilidad del método multivariante gráfico "biplot" a los estudios bibliométricos en biomedicina. in xiv jornadas nacionales de información y documentación en ciencias de la salud, cádiz, 13-15 de abril de 2011. (unpublished). Conference poster, http://hdl.handle.net/10760/15998.

Beh, E. J., & Lombardo, R. (2014). *Correspondence analysis, practice*. Methods and new strategies. Chichester: Wiley.

Beh, E. J., & Lombardo, R. (2021). *An introduction to correspondence analysis*. Chichester: Wiley.

Carlier, A., & Kroonenberg, P. M. (1996). Decompositions and biplots in three-way correspondence analysis. *Psychometrika, 61*, 355–373.

Ceyhun, E., Gokce, B., & Abdullah, Y. (2020). Economic policy responses to a pandemic: Developing the covid-19 economic stimulus index. *CEPR press, 3*, 40–53.

De Rooij, M., & Heiser, W. (2005). Graphical representations and odds ratios in a distance association model for the analysis of cross-classified data. *Psychometrika, 70*, 99–123.

Descartes, R. (1637). Discours de la méthode. Essay, Academia, Leiden.

Eckart, C., & Young, G. (1936). The approximation of one matrix by another of lower rank. *Psychometrika, 1*, 211–218.

Gabriel, K. R. (1971). The biplot graphic display of matrices with application to principal component analysis. *Biometrika, 58*, 453–467.

Gabriel, K. R. (1972). Analysis of meteorological data by means of canonical decomposition and biplots. *Journal of Applied Meteorology, 11*, 1071–1077.

Gabriel, K. R., & Odoroff, C. L. (1990). Biplots in biomedical research. *Statistics in Medicine, 9*, 469–485.

Galindo, P., Vaz, T., & Nijkamp, P. (2011). Institutional capacity to dynamically innovate: An application to the portuguese case. *Technological Forecasting and Social Change, 78*, 3–12.

Galindo, P. V. (1986). Una alternativa de representación simultánea: HJ biplot. *Questió, 10*, 13–23.

Goodman, L. A. (1979). Simple models for the analysis of association in cross-classifications having ordered categories. *Journal of the American Statistical Association, 74*, 537–552.

Goodman, L. A. (1985). The analysis of cross-classified data having ordered and/or unordered categories: Association models, correlation models and asymmetry models for contingency tables with or without missing entries. *The Annals of Statistics, 13*, 10–69.

Gower, J. C., & Hand, D. J. (1996). *Biplots*. Chapman and Hall.

Gower, J. C., & Harding, S. A. (1988). Nonlinear biplots. *Biometrika, 75*, 445–455.

Gower, J. C., Le Roux, N., & Lubbe, S. (2015). Biplots: Quantitative data. *WIREs Computational Statistics, 7*, 42–62.

Gower, J. C., Le Roux, N., & Lubbe, S. (2016). Biplots: Qualititative data. *WIREs Computational Statistics, 8*, 82–111.

Gower, J. C., Lubbe, S., & Le Roux, N. (2011). *Understanding biplots*. Chichester: Wiley.

Greenacre, M. (2013). Contribution biplots. *Journal of Computational and Graphical Statistics, 22*, 107–122.

Greenacre, M. J. (1984). *Theory and application of correspondence analysis*. London: Academic.

Greenacre, M. J. (1993a). Biplots in correspondence analysis. *Journal of Applied Statistics, 20*, 251–269.

Greenacre, M. J. (1993b). *Correspondence analysis in practice*. London: Academic.

Greenacre, M. J. (2010a). *Biplots in practice*. Barcelona: Foudation BBVA.

Greenacre, M. J. (2010b). Log-ratio analysis is a limiting case of correspondence analysis. *Mathematical Geosciences, 42*, 129–134.

Greenacre, M. J., & Lewi, P. (2009). Distributional equivalence and subcompositional coherence in the analysis of compositional data, contingency tables and ratio-scale measurements. *Journal of Classification, 26*, 29–54.

Groenen, P., Le Roux, N., & Gardner-Lubbe, S. (2015). Spline-based nonlinear biplots. *Advances in Data Analysis and Classification, 9*, 219–238.

Hotelling, H. (1933). Analysis of a complex of statistical variables into principal components. *Journal of Education Psychology, 24*, 498–520.

Husson, F., Le, S., & Pages, J. (2010). *Exploratory multivariate analysis by example using R*. Chapman and Hall.

Lewi, P. J. (1976). Spectral mapping, a technique for classifying biological activity profiles of chemical compounds. *Arzneimittel Forschung, 26*, 1295–1300.

Loach, J. (1998). Le Corbusier and the creative use of mathematics. *The British Journal for the History of Science, 31,* 185–215.

Lombardo, R., Carlier, A., & D'Ambra, L. (1996). Nonsymmetric correspondence analysis for three-way contingency tables. *Methodologica, 4,* 59–80.

Mardia, K. V., Kent, J. T., & Bibby, J. M. (1979). *Multivariate analysis.* New York: Academic.

Pan, S., Chon, K., & Song, H. (2008). Visualizing tourism trends: A combination of atlas.ti and biplot. *Journal of Travel Research, 46,* 339–348.

Pearson, K. (1901). On lines and planes of closest fit to a system of points in space. *Philosophical Magazine, 6,* 559–572.

Torres-Salinas, D., Robinson-García, N., Jiménez-Contreras, E., Herrera, F., & López-Cóza, E.D (2013). On the use of biplot analysis for multivariate bibliometric and scientific indicators. *Journal of the American Society for Information Science and Technology, 64,* 1468–1479.

Udina, F. (2005). Interactive biplot construction. *Journal of Statistical Software*, 13:16 pages.

Veiga de Cabo, J., & Martín-Rodero, H. (2011). Acceso abierto: nuevos modelos de edicion cien-tíÂfica en entornos web 2.0. *Salud Colectiva, 7,* 519–527.

Wouters, L., Gohlmann, H., Bijnens, L., Kass, S., Molenberghs, G., & Lewi, P. (2003). Graphical exploration of gene expression data: A comparative study of three multivariate methods. *Biometrics, 59,* 1131–1139.

Yan, W., & Kang, M. (2003). *GGE biplot analysis: A graphical tool for breeders, geneticists, and agronomists.* Boca Raton: CRC Press.

Yan, W. K., Hunt, L., Sheng, Q., & Szlavnics, Z. (2000). Cultivar evaluation and megaenvironment investigation based on the gge biplot. *Crop Science, 40,* 597–605.

Chapter 10
Biplots for Variants of Correspondence Analysis

10.1 Introduction

In the previous chapter, we gave an overview and application of biplots for numerical data. We described the three types of biplots that one may construct—the row isometric, column isometric, and symmetric biplots—and we demonstrated the utility of the first type by analyzing data from 15 countries around the world and their financial and fiscal response to the COVID-19 global pandemic.

In this chapter, we focus our attention on discussing the role of biplots for the analysis of categorical variables and do so by describing their role in the correspondence analysis (CA) of a two-way and multi-way contingency table (Beh and Lombardo 2014). In doing so, we first briefly describe the construction of biplots for the *simple* correspondence analysis (SCA) of a contingency table formed from the cross-classification of two nominal variables; the term *simple* refers to performing CA for only two categorical variables. Such a technique involves the singular value decomposition (SVD) of a transformed matrix. In fact, this transformed matrix and the role of SVD was extensively discussed from a reciprocal averaging perspective in Chap. 7. We then discuss how SCA can be adapted when a two-way contingency table consists of two ordered categorical variables. In this case, rather than using SVD, we instead reflect the structure of the ordered variables using a bivariate moment decomposition (BMD) of the transformed matrix. Typically, for the visual display of the association between categorical variables, correspondence analysis involves constructing a correspondence plot which has been a topic of some debate for decades (Carroll et al. 1986, 1987, 1989)—some of the issues of this debate have been described by Nishisato and Clavel (2003); Nishisato (2016) and in earlier chapters of this book. Much of the concern has been centred on the "joint" aspect of this dislay, where a single plot is used to visually depict the two configuration of points that exist in different spaces; one exists in a *row space*, while the other exists in a *column space*. Much of the criticism has focused on the interpretation of the distance between a row point and a column point. Therefore, we shall not be describing the construction of a correspondence plot in this chapter. Instead, we focus our

S. Nishisato et al., *Modern Quantification Theory*,
Behaviormetrics: Quantitative Approaches to Human Behavior 8,
https://doi.org/10.1007/978-981-16-2470-4_10

discussion on the construction of the biplot since the inner product between a set of standard coordinates and principal coordinates does have a distance interpretation. For a discussion on the construction of correspondence plots and biplots from a correspondence analysis perspective, see, for example, Greenacre (1984) and Beh and Lombardo (2014).

There are some similarities in how one can go about constructing a biplot for numerical variables (using principal components analysis (PCA)) and constructing a biplot for studying the association between categorical variables (using SCA). As we described from a PCA perspective in Chap. 9, the rows of the data matrix, say, can be viewed as a point in a low-dimensional plot by displaying their position using principal coordinates, while the columns of the matrix can be viewed as a projection from the origin to the point defined by their standard coordinates. Such a display is referred to as a row isometric biplot and we constructed such a visual display in the application of Chap. 9. However, a clear distinction between the analysis of numerical variables and categorical variables is that, from the cross-classification of two categorical variables, the rows of the contingency table are the categories of one variable, while its columns are the categories of the second variable. By constructing a contingency table, the data values are counts of how many individuals fall into each pairing of categories for the two variables.

We shall introduce the biplot by examining the association between two categorical variables using SCA. In doing so, we need to consider the way in which the two variables are associated, since this will define the measure of association that shall be used. In this chapter, the well-known Pearson chi-squared statistic (Pearson 1904) will be used when it is assumed, or known, that there exists a *symmetric* association between two categorical variables. In this case, all categorical variables are treated as being a predictor variable. For a two-way contingency table formed from two symmetrically associated variables, transposing the table will have no inpact on the nature of the association between the variables and the interpretation of the resulting biplot remains unchanged. The second type of association that can be considered is when the categorical variables have an *asymmetric association*. In this case, at least one categorical variable is treated as a response variable given that the remaining variables are defined as a predictor variable, and the measure of association that can be used is the Goodman-Kruskal tau index (Goodman and Kruskal 1954). These two measures of association can be generalized for studying the association between three (or even more) categorical variables. For the sake of simplicity, our examination on the association between multiple categorical variables will be restricted to the case of three variables. In doing so, when all three variables are symmetrically associated and are cross-classified to form a three-way contingency table, we can use the three-way Pearson chi-squared statistic (Lancaster 1953) as our measure of association. However, it is worth noting that three-way generalizations of the Goodman-Kruskal tau index do exist. Three such generalizations are the Marcotorchino index (Marcotorchino 1985), the Gray-Williams index (Gray and Williams 1981), and the Delta index (Lombardo 2011); we shall not discuss the role of biplots for studying the asymmetric association between multiple categorical variables. Instead, the interested reader is invited to peruse the pages of Beh and Lombardo (2014, 2020) for more information on these indices and their role in the construction of biplots for three categorical variables.

10.2 Biplots for Simple Correspondence Analysis—The Symmetric Case

Before portraying the association between two symmetrically associated categorical variables using a biplot, we first briefly describe SCA. Define a $m \times n$ two-way contingency table, $\mathbf{F}$, formed from the cross-classification of a row and a column categorical variable. Denote the (i, j)th cell entry by f_{ij} for $i = 1, 2, \ldots, m$ and $j = 1, 2, \ldots, n$. Let the grand total of $\mathbf{F}$ be f_t and define the matrix of relative frequencies, $\mathbf{P}$, so that the (i, j)th entry is $p_{ij} = f_{ij}/f_t$ where $\sum_{i=1}^{m}\sum_{j=1}^{n} p_{ij} = 1$. Denote the ith marginal proportion of the row variable by $p_{i\bullet} = \sum_{j=1}^{n} p_{ij}$ which is the (i, i)th element of the diagonal matrix $\mathbf{D}_r$. Similarly, define the jth marginal proportion of the column variable as $p_{\bullet j} = \sum_{i=1}^{m} p_{ij}$ which is the (j, j)th element of the diagonal matrix $\mathbf{D}_c$.

Let Π be the two-way table where the (i, j)th element is the Pearson residual

$$\pi_{ij} = \frac{p_{ij}}{p_{i\bullet}p_{\bullet j}} - 1 .$$

such that

$$\Pi = \mathbf{D}_r^{-1}\left(\mathbf{P} - \mathbf{r}_m\mathbf{c}_n^T\right)\mathbf{D}_c^{-1} \tag{10.1}$$

where $\mathbf{r}_m = (p_{1\bullet}, \ldots, p_{m\bullet})^T$ and $\mathbf{c}_n = (p_{\bullet 1}, \ldots, p_{\bullet n})^T$. Then Pearson's chi-squared statistic, X^2, can be written as

$$\begin{aligned} X^2 &= f_t \sum_{i=1}^{m}\sum_{j=1}^{n} p_{i\bullet}p_{\bullet j}\left(\frac{p_{ij}}{p_{i\bullet}p_{\bullet j}} - 1\right)^2 \\ &= f_t \sum_{i=1}^{m}\sum_{j=1}^{n} p_{i\bullet}p_{\bullet j}\pi_{ij}^2 . \end{aligned}$$

One way that SCA can be performed is by applying a generalized singular value decomposition (GSVD) on Π such that

$$\Pi = \mathbf{X}\Lambda\mathbf{Y}^T . \tag{10.2}$$

In Eq. (10.2), $\mathbf{X}$ and $\mathbf{Y}$ are $m \times K$ and $n \times K$ column matrices, respectively, containing the left and right singular vectors where $K = \min(m, n) - 1$. They have the property $\mathbf{X}^T \mathbf{D}_r \mathbf{X} = \mathbf{I}$ and $\mathbf{Y}^T \mathbf{D}_c \mathbf{Y} = \mathbf{I}$, respectively, while Λ is the $K \times K$ diagonal matrix containing the singular values of Π and are arranged in descending order.

The row and column categories of Π can then be visualized by constructing a row or column *low-dimensional biplot*; see Greenacre (1984), Greenacre (2007b), Gower et al. (2011), and Beh and Lombardo (2014, 2020) for further details. In Chap. 9, we described how to construct a biplot for examining the relationships that exist between numerical variables. A similar strategy can be adopted for studying the association between categorical variables.

For SCA, the association between the row variable and the column variable can be visualized using a row isometric biplot or a column isometric biplot. For the row isometric biplot, such a visualization can be constructed using the row and column coordinates

$$\mathbf{R} = \mathbf{X}\Lambda \tag{10.3}$$

and

$$\mathbf{C}^s = \mathbf{Y}\,, \tag{10.4}$$

respectively. Here $\mathbf{R}$ is the $m \times K$ column matrix containing the row principal coordinates while $\mathbf{C}^s$ is the $n \times K$ column matrix containing the column standard coordinates. Therefore, the biplot is constructed of at most K dimensions. Based on this definition of the biplot coordinates, the matrix of Pearson residuals can be expressed as the inner product of the coordinates such that $\Pi = \mathbf{R}\,(\mathbf{C}^s)^T$.

A column isometric biplot can be constructed by defining the row and column coordinates by

$$\mathbf{R}^s = \mathbf{X}$$

and

$$\mathbf{C} = \mathbf{Y}\Lambda\,,$$

respectively. For this biplot, $\mathbf{R}^s$ is the $m \times K$ column matrix that contains the row standard coordinates and the $n \times K$ column matrix $\mathbf{C}$ contains the column principal coordinates so that the optimal number of dimensions of this biplot is K. Based on this definition of the biplot coordinates their inner product gives the matrix of Pearson residuals so that $\Pi = \mathbf{R}^s\mathbf{C}^T$.

For these two isometric biplots, their quality can be determined to assess how much of the association is captured in two, or more dimensions. This assessment can be made by calculating the *total inertia*, ϕ^2, of the contingency table which is akin to Pearson's phi-squared statistic (Pearson 1904, p. 6) and is calculated by

$$\phi^2 = \frac{X^2}{f_t} = \text{trace}\left(\Lambda^2\right)\,.$$

The quality of a two-dimensional biplot, say, can then be determined by calculating the percentage contribution of the sum-of-squares of the first two singular values, $\lambda_1^2 + \lambda_2^2$, to ϕ^2; the larger $\lambda_1^2 + \lambda_2^2$, the better the quality of the display. For a comprehensive technical, practical, and historical description of categorical (qualitative) scaling and their link to SCA, the interested reader can refer to Nishisato (1980), Greenacre (1984), Beh and Lombardo (2012, 2014), and the references mentioned therein for further details.

As we described in the previous chapter, only within-variable (squared) distances between points that are depicted by their principal coordinates can be interpreted and are Euclidean (Beh and Lombardo 2014, p. 132). Although one may assess the strength of an interaction between two categories from different variables by assessing how close a point depicted by a principal coordinate lies from the projection made from the origin to a point defined by a standard coordinate.

While this section has provided an overview of SCA for nominal categorical variables, SCA can also be used to examine the association between two ordinal categorical variables. Such a technique can also be extended for the analysis of multiple categorical variables where are least one of them is ordinal. Performing such variants of CA requires alternative methods of decomposition, and we shall describe some of these in the coming sections. Before we do so, we now turn our attention to SCA, where the two categorical variables have an asymmetric association structure.

10.3 Biplots for Simple Correspondence Analysis—The Asymmetric Case

When studying the association between two categorical variables, it may be (by nature of the variables, or by assumption) that one of them is deemed a predictor variable and the other a response variable. In this case, the association is defined as being *asymmetric* in nature. In this case, Pearson's chi-squared statistic is not a suitable measure of association. Instead, one can use the Goodman-Kruskal tau index (Goodman and Kruskal 1954) to quantify this association. In some studies, this index is referred to as an *index of predictability* since it measures the relative increase in predictability of the row (response) categories given the column (predictor) categories and is defined by

$$\tau = \frac{\sum_{i=1}^{m}\sum_{j=1}^{n} p_{\bullet j}\left(\frac{p_{ij}}{p_{\bullet j}} - p_{i\bullet}\right)^2}{1 - \sum_{i=1}^{m} p_{i\bullet}^2} .$$

The index τ is by 0 and 1 (inclusive), where a value of 1 reflects perfect predictability of the rows categories given the columns, while a value of 0 reflects no prediction (and coincides with complete independence between the variables). Since the denominator of this index is independent of cell counts/proportions, the correspondence analysis literature concerned with two categorical variables having a predictor/response association structure quantifies the total inertia using only the numerator of the Goodman-Kruskal tau index. In this case, the variant of correspondence analysis is called *non-symmetrical correspondence analysis* (NSCA) and has been described in quite detail since its development (D'Ambra and Lauro 1989; Lauro and D'Ambra 1984) by Kroonenberg and Lombardo (1999), Beh and Lombardo (2014), and Beh and Lombardo (2020). Therefore, by using this numerator of τ, denoted by τ_{num} as the total inertia, NSCA can be performed by first defining the matrix

$$\Pi_\tau = \left(\mathbf{P} - \mathbf{r}_m \mathbf{c}_n^T\right) \mathbf{D}_c^{-1} ,$$

which is just the matrix of the weighted centred column profiles. NSCA is then undertaken by applying a generalized singular value decomposition (GSVD) on Π_τ so that

$$\Pi_\tau = \tilde{\mathbf{X}} \Lambda \mathbf{Y}^T . \tag{10.5}$$

where the left and right singular vectors, $\tilde{\mathbf{X}}$ and $\mathbf{Y}$, have the property $\tilde{\mathbf{X}}^T \tilde{\mathbf{X}} = \mathbf{I}$ and $\mathbf{Y}^T \mathbf{D}_c \mathbf{Y} = \mathbf{I}$, respectively.

The asymmetric association between the row (response) and column (predictor) variables can then be visualized from the matrices of this GSVD by constructing a *column isometric biplot*. Such a biplot is constructed using the row standard coordinates $\mathbf{R}^s = \tilde{\mathbf{X}}$ and the column principal coordinates $\mathbf{C} = \mathbf{Y}\Lambda$. If one were interested in defining the row variable as being predictor variable, so that the column variable is the response variable, then a row isometric biplot is an appropriate means of visualizing the asymmetric association between the variables. For these biplots, their quality can assess how much of the predictability is captured in two or more dimensions by calculating the *total inertia*, τ_{num}, of the contingency table which is the numerator of Goodman-Kruskal statistic (Goodman and Kruskal 1954), such as

$$\tau_{\text{num}} = \text{trace}\left(\Lambda^2\right) .$$

For more information on the role of biplots in NSCA, see Kroonenberg and Lombardo (1999) and Beh and Lombardo (2020) for further details.

10.4 Ordered Simple Correspondence Analysis

10.4.1 An Overview

When the row and column variables both consist of ordered categories, this ordered structure often needs to be incorporated into the CA of the two-way contingency table. This can be achieved by reflecting the ordered structure by defining a priori scores that reflect this structure for the row categories and for the column categories. Using a priori scores is very common for performing a variety of different categorical data analysis techniques; see, for example, Agresti (2007, 2013) for a general discussion of this issue. In doing so, we define $s_r(i)$ to be the score assigned to the ith row category and $s_c(j)$ to be the score assigned to the jth column category. A special case of these scores are *natural scores*, where for the m row category $s_r(i) = i$, for $i = 1, 2, \ldots, m$. Similarly, the natural scores for the n column categories are defined by $s_c(j) = j$, for $j = 1, 2, \ldots, n$. Such scores imply that the ordering of both sets of categories is increasing and that the categories are equi-distant from one another. They also provide some very simple but intepretable, numerical, and visual summaries to be obtained when performing CA. Just like we saw when we discussed the role of reciprocal averaging in the previous chapters, the resulting scores are standardized so that they are centred at zero and have a variance of 1. For a priori scores, this standardization procedure can be undertaken using the scores as a basis vector and employing the Gram-Schmidt orthogonalization procedure. A much simpler way of obtaining such a set of standardized scores from the a priori scores is to use the simple recurrence formulae to generate orthogonal polynomials. As we have often done in the past, here we use the formulae described by Emerson (1968). More information on the link between the Gram-Schmidt process and Emerson's (1968) recurrence formulae was described in Chap. 3 of Beh and Lombardo (2014).

It is important at this point to highlight that this is where the scoring procedures for nominal categorical variables and those for ordinal categorical variables diverge. As we have described in previous chapters of this book, there are a variety of ways in which scores for nominal categories can be determined, all of which have been described from a reciprocal averaging perspective. Certainly, if reciprocal averaging yields a set of (one-dimensional) ordered scores that reasonably reflect the ordered structure of the variable, then these can be used as a basis for constructing orthogonal polynomials; see Beh (1998) for details on the impact of doing so when performing an ordered correspondence analysis on a two-way contingency table. Beh (1998) also examined the use of the scores described by Nishisato and Arri (1975), who derived a procedure for determining scores for ordered categorical variables, but where one category did not follow the order (such as a "Don't Know" category).

So, suppose we incorporate the structure of two ordered categorical and symmetrically associated variables using a priori scores that are standardized using orthogonal polynomials generated using Emerson's (1968) formulae. The analyst can reflect this structure in the CA of the contingency table by applying a *bivariate moment decomposition* (BMD) to the table of Pearson's ratios—see Eq. (10.1). Mathematically, BMD looks quite similar in structure to SVD and is of the form

$$\Pi = \mathring{\mathbf{X}}\mathring{\Lambda}\mathring{\mathbf{Y}}^T \tag{10.6}$$

where $\mathring{\mathbf{X}}^T\mathbf{D}_r\mathring{\mathbf{X}} = \mathbf{I}$ and $\mathring{\mathbf{Y}}^T\mathbf{D}_c\mathring{\mathbf{Y}} = \mathbf{I}$. Here $\mathring{\mathbf{X}}$ is the column matrix of orthogonal polynomials and is of size $m \times (m-1)$ while $\mathring{\mathbf{Y}}$ is the column matrix of column orthogonal polynomials and is of size $n \times (n-1)$. The first column of each matrix reflects polynomials that are linearly arranged in the same manner as how the a priori scores are defined, while their second column is arranged in a cubic manner, and so on. The matrix $\mathring{\Lambda}$ contains the generalized correlations between $\mathring{\mathbf{X}}$ and $\mathring{\mathbf{Y}}$. Therefore, unlike the matrix of singular values, Λ, obtained from the SCA of two nominal categorical variables, $\mathring{\Lambda}$ is not a diagonal matrix and is a bi-orthogonal rectangular matrix of size $(m-1) \times (n-1)$. Its elements reflect various sources of linear and nonlinear association that exist between the row and column variables. For example, the $(1,\ 1)$th element is the linear-by-linear correlation between the ordered row and column variables. Depending on the choice of scores, this correlation term is equivalent to Pearson's product moment correlation (when natural scores are used for both the row and column categories) or Spearman's rank correlation (when midrank scores are used). Similarly, the $(2,\ 1)$th element is the quadratic-by-linear association and is one source of nonlinear association between the categories that summarizes the association in terms of the dispersion of the row categories and the location (mean) of the column categories.

To get a clearer indication of the structure of this method of decomposition, suppose we consider the $(i,\ j)$th element of Π

$$\pi_{ij} = \sum_{u=1}^{m-1} \sum_{v=1}^{n-1} \mathring{x}_{iu} \mathring{\lambda}_{uv} \mathring{y}_{jv} .$$

Here $\mathring{x}_{iu}$ is a uth order orthogonal polynomial for the ith row category, while $\mathring{y}_{jv}$ is a vth order orthogonal polynomial for the jth column category; these terms are the $(i,\ u)$th and $(j,\ v)$th elements of $\mathring{\mathbf{X}}$ and $\mathring{\mathbf{Y}}$ respectively, for $u = 1, \ldots, m-1$ and $v = 1, \ldots n-1$. The term $\mathring{\lambda}_{uv}$ is the $(u,\ v)$th element of $\mathring{\Lambda}$ and may also be interpreted as the correlation between the uth order row polynomial and the vth order column polynomial so that

$$\mathring{\lambda}_{uv} = \sum_{u=1}^{m-1} \sum_{v=1}^{n-1} p_{ij} \mathring{x}_{iu} \mathring{y}_{jv} . \tag{10.7}$$

An important feature of applying a BMD to Π is that the total inertia of the contingency table can be expressed as the sum-of-squares of the generalized correlations so that

$$\frac{X^2}{f_t} = \sum_{u=1}^{m-1} \sum_{v=1}^{n-1} \mathring{\lambda}_{uv}^2 . \tag{10.8}$$

This total inertia may be expressed in matrix form by

$$\frac{X^2}{f_t} = \text{trace}\left(\mathring{\Lambda}^T \mathring{\Lambda}\right) = \text{trace}\left(\mathring{\Lambda} \mathring{\Lambda}^T\right) .$$

Expressing the chi-squared statistic (or the total inertia) in this manner has been described by Lancaster (1953), Best and Rayner (1996), Rayner and Best (1996), Beh (1997), and many others. Further details and variations of this particular partition of Pearson's chi-squared statistic on this can be found by referring to, for example, Rayner and Best (2001), Beh and Davy (1998, 1999), and Beh (2001).

This variant of correspondence analysis has been extensively described and elaborated upon since the 1990s, with the aim of accommodating for the structure of ordered categorical variables; see, for example Beh (1997, 1998, 2004), Lombardo et al. (2007), Lombardo and Beh (2010) Beh and Lombardo (2014), Lombardo and Meulman (2010), Simonetti et al. (2011), and Lombardo et al. (2016) for additional descriptions of their use in this and related context.

10.4.2 Biplots for Ordered Simple Correspondence Analysis

Now that we have described how orthogonal polynomials and generalized correlations can be incorporated into the CA of a two-way contingency table with an ordinal row variable and an ordinal column variable, we discuss how one may construct a biplot to visually inspect the association between these variables. By applying a BMD to the matrix of Pearson residuals's, Π, the row and column *scores* are summarized in the columns of the matrix of row orthogonal polynomials, $\mathring{\mathbf{X}}$ and the matrix of column orthogonal polynomials, $\mathring{\mathbf{Y}}$, respectively. Unlike the calculation of such scores for nominal variables, which rests on techniques such as reciprocal averaging (see Chap. 7), these matrices can be used with the matrix of generalized correlations to create the three types of biplots we described in Chap. 9. We now turn our attention to describing how these biplots can be constructed.

To construct a row isometric biplot that visually displays the association between an ordered row variable and an ordered column variable of a two-way contingency table, the following coordinates can be defined

$$\mathring{\mathbf{R}} = \mathring{\mathbf{X}}\mathring{\Lambda} \quad (10.9)$$

$$\mathring{\mathbf{C}}^s = \mathring{\mathbf{Y}} \quad (10.10)$$

so that $\Pi = \mathring{\mathbf{R}}\left(\mathring{\mathbf{C}}^s\right)^T$. Here, $\mathring{\mathbf{R}}$ is a $m \times (n-1)$ column matrix of the row principal coordinates and $\mathring{\mathbf{C}}^s$ is a $n \times (n-1)$ column matrix containing the column standard coordinates. Therefore, this biplot is constructed so that the rows are depicted using points defined by their principal coordinates in a low-dimensional space, while the columns are depicted as a projection from the origin to the point defined by their standard coordinates in the same space. The maximum number of dimensions for this variant of the row isometric biplot is $n-1$ and not $K = \min(m, n) - 1$ as we saw for the joint visualization of the categories of a nominal row and a nominal column variable.

A column isometric biplot can also be constructed. To do so, the row standard and column principal coordinates are defined by the column matrices

$$\mathring{\mathbf{R}}^s = \mathring{\mathbf{X}} \tag{10.11}$$

$$\mathring{\mathbf{C}} = \mathring{\mathbf{Y}}\mathring{\Lambda}^T, \tag{10.12}$$

respectively. Therefore, from the BMD of Π—see Eq. (10.6)—this matrix of Pearson residuals's may be reconstituted from the inner product of these coordinates such that $\Pi = \mathring{\mathbf{R}}^s\mathring{\mathbf{C}}^T$. A full reconstitution can be achieved when $\mathring{\mathbf{R}}^s$ is of size $m \times (m-1)$ and $\mathring{\mathbf{C}}$ is of size $n \times (m-1)$. For the joint representation of these coordinates in the same space, the maximum dimensionality of this space is $m-1$ and can be constructed by depicting the columns using their principal coordinates and the rows as a projection from the origin to their position defined by their standard coordinates.

These two isometric biplots can provide an interpretation of those principal coordinates that share a similar (or different) position by calculating the squared Euclidean distance between their points. A small Euclidean distance between two principal coordinates provides some evidence that their categories contribute to the association structure between the variables in a similar manner. A large Euclidean distance suggests that their contribution to this association is different. One may also determine the *interaction* between a row point and a column point by observing the proximity of a principal coordinate from the projection made from the origin to a standard coordinate. A small distance between this point and the projection indicates that there is a strong interaction between the row/column pair, while a large distance suggests that their interaction is quite weak; such a distance can also be reflected by observing the sign and magnitude of the inner product between the standard and principal coordinates (Beh and Lombardo 2020).

Suppose we have a row isometric biplot. The total inertia can be expressed in terms of the row principal coordinates such that

$$\frac{X^2}{f_t} = \text{trace}\left(\mathring{\mathbf{R}}^T\mathbf{D}_r\mathring{\mathbf{R}}\right)$$

while for the column isometric biplot, the total inertia can be expressed in terms of the column principal coordinates by

$$\frac{X^2}{f_t} = \text{trace}\left(\mathring{\mathbf{C}}^T\mathbf{D}_c\mathring{\mathbf{C}}\right).$$

These expressions may be expressed using sigma notation such that

$$\frac{X^2}{f_t} = \sum_{v=1}^{n-1}\sum_{i=1}^{m} p_{i\bullet}\mathring{r}_{iv}^2 \tag{10.13}$$

and

$$\frac{X^2}{f_t} = \sum_{u=1}^{m-1}\sum_{j=1}^{n} p_{\bullet j}\mathring{c}_{ju}^2 \tag{10.14}$$

respectively, where $\mathring{r}_{iv}$ is the (i, v)th element of $\mathring{\mathbf{R}}$ (see Eq. (10.9)) and $\mathring{c}_{ju}$ is the (j, u)th element of $\mathring{\mathbf{C}}$ (see Eq. (10.12)). Therefore, principal coordinates that lie close to the origin reflect those categories that are not dominant contributors to the association, while principal coordinates that lie far from the origin reflect those categories that are dominant contributors.

An important feature of the principal coordinates is the interpretation of their distance from each other. We first note that the squared difference between the ith and i'th row profile is

$$d_I^2(i, i') = \sum_{j=1}^{n} \frac{1}{p_{\bullet j}} \left(\frac{p_{ij}}{p_{i\bullet}} - \frac{p_{i'j}}{p_{i'\bullet}} \right)^2$$
$$= \sum_{j=1}^{n} \frac{1}{p_{\bullet j}} \left[\left(\frac{p_{ij}}{p_{i\bullet}} - p_{\bullet j} \right) - \left(\frac{p_{i'j}}{p_{i'\bullet}} - p_{\bullet j} \right) \right]^2 .$$

When studying the association between two ordinal categorical variables using simple correspondence analysis, Beh (1997) showed that this result is equivalent to the squared Euclidean distance of their principal components in an optimal $(n - 1)$ dimensional space such that

$$d_I^2(i, i') = \sum_{v=1}^{n-1} \left(\mathring{r}_{iv} - \mathring{r}_{i'v} \right)^2 .$$

While Beh (1997) was concerned with these distances in a traditional correspondence plot (which jointly depicts the principal coordinates of ordered row and column categories) this result also applies to the row principal coordinates of a row isometric biplot.

These squared Euclidean distance results show that any differences between two row (or column) profiles is reflected by the squared distance between their principal coordinates in a row (or column) isometric biplot. Such a property thereby abides by the *property of distributional equivalence* described by Lebart et al. (1984)[p. 35], Greenacre (1984)[p. 95] and Beh and Lombardo (2014)[Sect. 4.6].

We can also quantify the squared distance of a (ordered) row principal coordinate from the origin of a row isometric biplot by

$$d_I^2(i, 0) = \sum_{j=1}^{n} \frac{1}{p_{\bullet j}} \left(\frac{p_{ij}}{p_{i\bullet}} - p_{\bullet j} \right)^2$$
$$= \sum_{v=1}^{n-1} \mathring{r}_{iv}^2 .$$

Therefore, from Eq. (10.13), the total inertia of the contingency table with an ordered row and column variable can be expressed as the weighted sum-of-squares of this squared distance since

$$\frac{X^2}{f_t} = \sum_{i=1}^{m} p_{i\bullet} d_I^2 \left(i,\ 0\right)$$

and one can draw the same conclusions about the proximity of a row point from the origin that we made when discussing the implications of Eqs. (10.13) and (10.14).

While we have confined our attention here to the features of the row principal coordinates in a row isometric biplot, similar remarks can also be made for the column principal coordinates of a column isometric biplot.

10.4.3 The Biplot and a Re-Examination of Table 3.1

Consider Table 3.1 that was introduced in Chap. 3, and studied by Nishisato (1980); and reproduced here as Table 10.1.

Recall that it is formed from the cross-classification of 140 individuals according to their *Propensity* to use sleeping pills and the frequency of *Nightmares* they have had when sleeping. The categories of the *Propensity* variable are as follows: *Strongly against*, *Against*, *Neutral*, *For*, and *Strongly For*. One can also see that the *Nightmare* variable consists of the five categories *Never*, *Rarely*, *Sometimes*, *Often*, and *Always*. Since both variables are ordered, we can perform an ordered simple correspondence analysis on Table 10.1. However, before doing so it is worth determining first whether there exists a statistically significant association between the *Propensity* and *Nightmare* variables. By performing a chi-squared test of independence on the contingency table, we find that Pearson's chi-squared statistic is $X^2 = 78.4$ with 16 degrees of freedom. Therefore, with a p-value that is less than 0.0001 there exists a statistically significant association between the two ordered variables. Note that it is

Table 10.1 Contingency Table of *Propensity* to take sleeping pills and frequency of *Nightmares*

Nightmares						
Propensity	Never	Rarely	Sometimes	Often	Always	Total
Strongly against	15	8	3	2	0	28
Agaist	5	17	4	0	2	28
Neutral	6	13	4	3	2	28
For	0	7	7	5	9	28
Strongly For	1	2	6	3	16	28
Total	27	47	24	13	29	140

not necessary to incorporate the "orderedness" of the row and column categories of Table 10.1 to determine the chi-squared statistic. However, if one wished to investigate further the nature of this association by partitioning X^2 then the structure of the ordered set of row and column categories can be incorporated using the a priori scores; see Eq. (10.8).

By applying a BMD to Π additional insight into the variation of the row and column profiles can be obtained by partitioning Pearson's chi-squared statistic into location, dispersion, and higher-order effects; for ordered SCA these are the inertia values for each dimension of the isometric biplot. Table 10.3 summarizes the variation that exists between the row profiles in terms of differences in their location (mean) dispersion (akin to standard deviation), cubic (akin to skewness), and the aggregation of the higher (labelled as "Error") components. Generally, the inertia value of the vth order column effect is

$$\mathring{\lambda}^2_{\bullet v} = n \sum_{u=1}^{n-1} \mathring{\lambda}^2_{uv} .$$

and is a chi-squared random variable with $m - 1$ degrees of freedom. Table 10.2 shows that the variation in the column profiles of Table 10.1 can be best described by the differences in their location (or mean) since the column location inertia is 55.11 and accounts for 70.53% of the association; one can also see that its p-value is very small thereby showing that there exists a statistically significant difference in the mean of each column profile. That is, the variation in the mean of their profiles is a dominant and a statistically significant source of the association between the variables. The variation in the dispersion of the column profiles is also statistically significant but not quite as dominant a source only accounting for about 17% of the association that exists between the variables of Table 10.1. The cubic effect reflects the variation in the skewness of the column profiles. Its large p-value suggests that this effect is not statistically significant. Nor is the "Error" effect which, in this case, reflects the variation in what may be described as the kurtosis of the column profiles.

Similar comments can also be made about sources of variation that exist between the row profiles. Suppose we defined

$$\mathring{\lambda}^2_{u\bullet} = n \sum_{v=1}^{m-1} \mathring{\lambda}^2_{uv} .$$

Table 10.2 The row effects, $\mathring{\lambda}^2_{u\bullet}$, from the ordered SCA of Table 10.1

Effect	Inertia value	% Inertia	P-value
Location	55.11	70.54	<0.01
Dispersion	13.42	17.17	0.01
Cubic	5.75	7.36	0.22
Error	3.85	4.93	0.43
X^2	78.13	100.00	<0.01

Table 10.3 The column effect, $\mathring{\lambda}^2_{\bullet v}$, from the ordered SCA of Table 10.1

Effect	Inertia Value	% Inertia	P-value
Location	57.39	73.46	<0.01
Dispersion	9.50	12.16	0.05
Cubic	8.73	11.17	0.07
Error	2.51	3.21	0.64
X^2	78.13	100.00	<0.01

to be the uth order row effect. Table 10.3 summarizes these values and shows that nearly three quarters (73.46%) of this variation can be accounted for by their difference in their mean, while their dispersion effect is not as dominant (12.16%.)

Further information about the association between the variables of Table 10.1 can also be obtained by calculating the generalized correlation values, λ_{uv}, for $u = 1, 2, 3, 4$ and $v = 1, 2, 3, 4$. Doing so reveals that the product moment correlation is $\lambda_{11} = 0.62$ and shows that there is a strong, and dominant, linear-by-linear association between the *Propensity* to use sleeping pills and the frequency of *Nightmares* they have had when sleeping. We also find that $\lambda_{22} = 0.24$ which shows that there is a quadratic-by-quadratic association between these variables. That is, the dispersion of the row profiles impacts upon the dispersion of the column profiles.

In our ordered SCA of Table 10.1, we have defined the set of a priori row scores for the ordered *Propensity* categories to be the Likert scores described in Chap. 3 (being $-2, -1, 0, 1$, and 2). We use the same scores for the five ordered categories of the *Nightmare* variable. In fact, any linear transformation of a set of a priori scores can be chosen and the orthogonal polynomials, and hence the features from which they are based (including principal coordinates and the biplot) will remain unchanged. For example, one may have chosen natural scores for the rows and columns and the results from performing an ordered SCA will be identical to those obtained using the scores used in Chap. 3. For more on this see Beh (1998, Sect. 4.1.1.).

Suppose we perform the more traditional SCA approach described in Sect. 10.2 to Table 10.1. By doing so, we are (for the moment) ignoring the fact that the row and column categories are ordered and so we are treating them as being nominal in nature. For this analysis, the association between the *Propensity* and *Nightmare* variables can be visually summarized by a row isometric biplot; see Fig. 10.1. The optimal dimensionality of the biplot is $K = \min(5, 5) - 1 = 4$ while Fig. 10.1 depicts the first two of these dimensions.

Since Fig. 10.1 is a row isometric biplot, it has been constructed by using row principal coordinates and columns standard coordinates; see Eqs. (10.3) and (10.4), respectively. The quality of this plot can be assessed by considering the percentage contribution that each dimension makes to the total inertia. From Eq. (10.13), the percentage contribution of the v'th dimension (for $v = 1, 2, 3, 4$) to the total inertia is

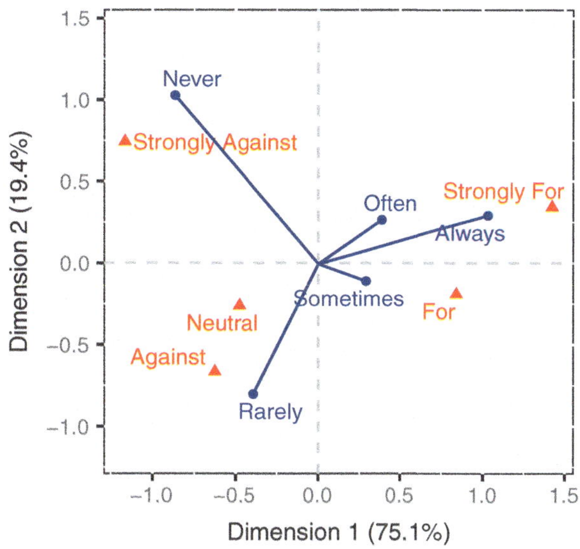

Fig. 10.1 Row isometric biplot from the simple correspondence analysis of Table 10.1

$$100 \times \frac{\sum_{i=1}^{5} p_{i\bullet} r_{iv}^2}{X^2/n}.$$

Alternatively, one may also quantify these values using the inertia values summarized in Table 10.3. Therefore, the first two dimensions visually depict 73.5% and 12.2% of the association between the variables, respectively.

Therefore, while a four-dimensional biplot will visually describe 100% of the association, the two-dimensional biplot given by Fig. 10.1 visually describes 94.5% of the association that exists between the variables. It is therefore deemed to be an excellent visual summary of the association between the variables of Table 10.1. Taking advantage of this quality, we can reliably see that there is a strong interaction between the category pairs *Strongly against* and *Never*, *Strongly For* and *Always*, and *Against* and *Rarely*, thereby agreeing with the visualization made by Fig. 10.1. Figure 10.1 also shows that there is a strong association between *For* and *Always*.

Similarly, for the column isometric biplot and based on Eq. (10.14), the percentage contribution that the uth dimension (for Table 10.1, $u = 1, 2, 3, 4$), to the total inertia is

$$100 \times \frac{\sum_{j=1}^{5} p_{\bullet j} c_{ju}^2}{X^2/n}$$

and these values are equivalent to the percentage inertias summarized in Table 10.2. So in Fig. 10.3, 70.5% of the association is depicted along the first dimension and 17.2% is depicted along the second dimension. Therefore, this column isometric biplot is a very good quality visual summary depicting 87.7% of the association that exists between the two variables of Table 10.1.

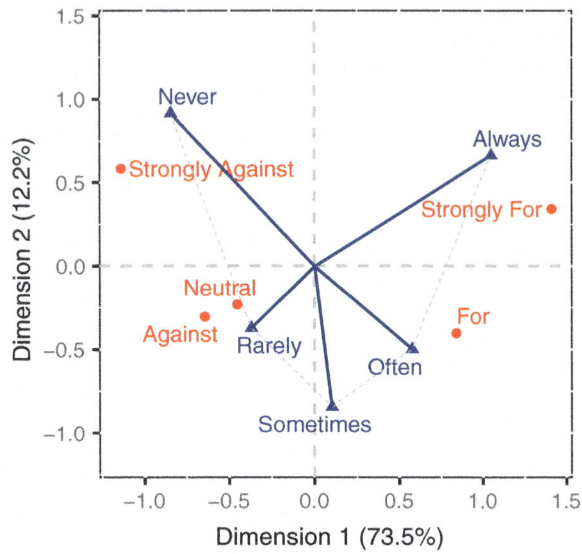

Fig. 10.2 Row isometric biplot from the ordered correspondence analysis of Table 10.1

Since the *Propensity* and *Nightmare* variables of Table 10.1 consist of ordered categories, we now incorporate this structure into the analysis using Likert scales $(-2, -1, 0, 1, 2)$ as the set of a priori scores for both variables and using them to generate orthogonal polynomials. We therefore, perform the ordered simple SCA outlined in Sect. 10.4 and construct a row isometric biplot that is given by Fig. 10.2. This biplot was constructed using the row principal coordinates of Eq. (10.9) and the column standard coordinates of Eq. (10.10). For the sake of comparison, we also produce the column isometric biplot—see Fig. 10.3—which has been constructed using the row standard coordinates of Eq. (10.11) and the column principal coordinates of Eq. (10.12). Since the standard coordinates are defined using orthogonal polynomials, their configuration in the biplots reflects the first two orders of those polynomials. That is, the standard coordinates are arranged along the first dimension in a linear fashion that reflects the ascending order of the scores chosen. For example, for the construction of the row isometric biplot of Fig. 10.2, the column categories are arranged in the same order they appear in the contingency table if one moves from the left hand side to the right hand side. The configuration of the standard coordinates along the second dimension of the biplots reflects the quadratic nature of the second order orthogonal polynomial and this can be seen by observing the parabolic configuration of the standard coordinates in Fig. 10.2 and Fig. 10.3. Note that the standard coordinates are also depicted as a projection from the origin to their position in the biplot. Therefore, an interpretation of the association between row and column categories in these isometric biplots can be made by observing the proximity of the principal coordinates from these projections.

Suppose we consider the row isometric biplot of Fig. 10.2 and observe the configuration of points along the first dimension. By assuming that there exists a symmetric

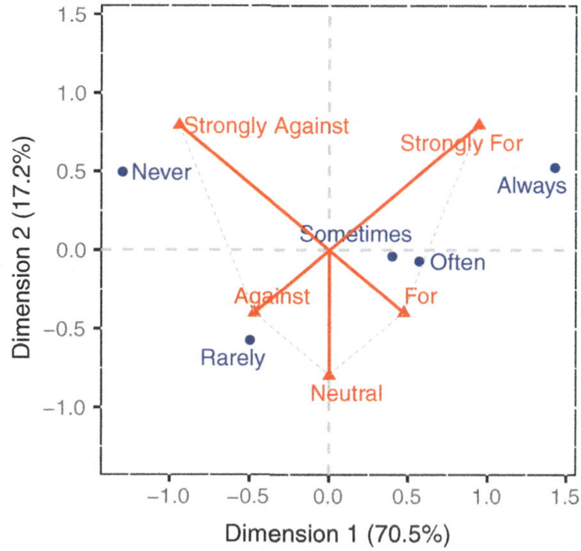

Fig. 10.3 Colum isometric biplot from the ordered correspondence analysis of Table 10.1

association between the row and column variables of Table 10.1, a simple interpretation would be that the *Propensity* to use sleeping pills is a linear function of the frequency of *Nightmare*. That is, the more sleeping pills one takes, the more likely that a person is to experience nightmares; indeed, this is what Fig. 3.1 also shows. The left side of the isometric biplot of Fig. 10.2 shows that there is a strong link between people who are *Strongly against* taking sleeping pills and those who *Never* experience nightmares. There is an even stronger link between those who are *Against* or remain *Neutral* to taking the pills and those who *Rarely* experience nightmares. On the right side of Fig. 10.2, we see that there is a strong link between those who are *Strongly For* taking sleeping pills and those who *Always* experience nightmares. It also appears that those who are *For* taking sleeping pills *Often* experience nightmares. However, it appears that those who *Sometimes* take sleeping pills are not strongly associated with any particular attitude to taking sleeping pills. Indeed, it seems that such individuals are most likely to be a mix of those who are either *Against* to *For* taking sleeping pills. In fact, if we observe the position of *Sometimes* in the column isometric biplot—see Fig. 10.3—we see that this column category is closest to the origin than any other column category. These conclusions can be observed by noting the proximity of the principal coordinates in the biplots and they are, generally, consistent with what Fig. 3.1 shows. However, it is unclear from Fig. 3.1 which (if any) row and column category of Table 3.1 plays a dominant role in how the association between the two categorical variables is defined. Such a feature can be identified by observing the proximity of the principal coordinates from the origin of the biplots. Also, while Fig. 3.1 shows a fairly weak, but apparent, association between *Sometimes* and *Neutral*, such a finding is not observed in the column isometric biplot.

Continuing to discuss the configuration of points in Figs. 10.2 and 10.3, since the relationship between the ordered variables is symmetric, we could also treat the frequency of *Nightmare* as a linear function of the *Propensity* to use sleeping pills. The reason for such a configuration is due to the linear structure of the first-order orthogonal polynomial. Since the scores we used to define the "orderedness" of the row and column categories are increasing (and equi-distant), the left hand side of the isometric biplots show those categories that exhibit a negative mean profile. That is, if we weight the elements of the centred row and column profiles, the mean profile of *Strongly against*, *Neutral*, and *Against* will be less than the "average" row profile (centred at the origin) suggesting that those row categories will be influenced more by the smaller scores than the larger scores. Similarly, since *Strongly For* and *For* lie on the right hand side of Fig. 10.2, they will have a positive mean profile which shows these row categories are dominated more by the higher scores than the lower scores. The configuration along the second dimension of the isometric biplots reveals a more complex association that exists between the variables. The second order polynomial reflects variation in the dispersion of the profiles, akin to (but not being equivalent to) the profiles standard deviation. So, by observing the configuration of principal coordinates along the second dimension of Fig. 10.2, it shows that while *Strongly against* and *Strongly For* have very different mean profiles, the variation exhibited between their profiles is very similar; this is because their position along the second dimension is very similar. Figure 10.2 also shows that, while *Strongly For* and *For* have a similar mean profile, their dispersion is very different; this is because, while they are positioned fairly closely along the first dimension, they have a very different position along the second dimension. Simple figures, like Fig. 3.1, clearly show variations in the mean profile of the categories and the strong linear-by-linear association that exists between the variables but, unlike the isometric biplots contructred from performing an ordered SCA, it neglects any nonlinear sources of association that may exist between the variables.

While the standard coordinates in both Figs. 10.2 and 10.3 have a parabolic configuration, the two configurations are not identical. Although it has not been demonstrated here, the second order polynomial has a perfect quadratic relationship with the first-order linear polynomial and so any configuration of standard coordinates will be perfectly parabolic. However, the two parabolic shapes have different characteristics, including the vertex, focus, directrix, and intercepts. Interestingly, while orthogonal polynomials have been used extensively in the CA literature since the late 1990s (and even longer within the statistical and her allied disciplines), very little attention has been given to these characteristics. So there is plenty of room to explore the geometric and numerical features of this area of ordered SCA.

While these biplots may be inferior in quality to the row isometric biplot when compared with Fig. 10.1, they do provide additional insight into exactly how the row and column variables are associated by observing variations in the profiles mean and dispersion. Also, unlike Fig. 10.1, using ordered SCA to construct an isometric biplot ensures that the ordered structure of both variables is incorporated into the analysis.

The two-way nominal and ordered correspondence analysis has been carried out using the R package `CAvariants` (Lombardo and Beh 2016).

10.5 The Biplot for Multi-Way Correspondence Analysis

10.5.1 An Overview

When analyzing the association between more than two variables, we can adopt the strategy that Kroonenberg (2008) referred to as "flattening" where a multi-way contingency table is transformed into a two-way table. Such a strategy is the most common approach that is adopted for performing a correspondence analysis and is referred to as the method of *multiple correspondence analysis* (MCA) (Greenacre 1984; Beh and Lombardo 2014). There are three ways in which 'flattening" can be done. A multi-way contingency table can be transformed into an indicator matrix of 0's and 1's; a 1 indicates that the individual is observed to have the characteristic described by a category, otherwise it is zero. A second approach is to transform the contingency table into its Burt matrix form which is a super-diagonal matrix of univariate and bivariate marginal frequencies. These two approaches are certainly the most common strategies for performing MCA and have been extensively discussed in the literature. Lastly, "flattening" can arise by stacking which involves placing each slice (or two-way table) of the three-way (say) contingency table on top of each other; see, for example, Weller and Romney (1990), Konig (2010), and Kim and Grochowalski (2019) for descriptions and applications of this approach.

A problem with these approaches is that, strictly speaking, they only capture two-way associations for pairs of variables and so do not properly incorporate any three-way (or higher-way) association structures that may exist in the contingency table. Flattening a multi-way contingency table "leads to an unnecessary and sometimes unacceptable simplification" (Kroonenberg 2008, p.7). An alternative approach is to analyze the contingency table as a *hypercube*, thereby performing *multi-way correspondence analysis* (MWCA) (Carlier and Kroonenberg 1996); one may refer to, for example, Kroonenberg (2008) and Beh and Lombardo (2019) for a description of MWCA. We now provide an overview of MWCA for the analysis of three categorical variables and describe the role that the biplot plays to visually assess the association between them.

10.5.2 TUCKER3 Decomposition

While CA (and MCA) involves decomposing a transformation of the data using SVD, when the cube/hypercube structure of the table is to be preserved, one needs to consider a multivariate version of SVD. In the correspondence analysis, and related, literature two such extensions of SVD that have proven popular are the CANDECOMP ("CANonical DECOMPosition") and the PARAFAC ("PARAllel FACtor") methods (Harshman 1970; Carroll and Chang 1970); while being independently proposed in 1970 they have been shown to yield equivalent score and correlation solutions. A second method of decomposition, and one that we shall consider in the remainder of this chapter, is the Tucker3 decomposition (Tucker 1964, 1966; Kroonenberg and de Leeuw 1980); which is designed for a three-way contingency table, but can be extended to analyze the association between multiple categorical variables.

To describe the Tucker3 decomposition, we turn our attention to describing the notation required. One may adopt the notation and terminology outlined in Kiers (2000), who wished to address the inconsistent vernacular and notation used in the development of the PARAFAC and CANDECOMP decomposition methods. However, we shall adopt a more conventional set of notation used in the categorical data analysis literature (with some minor differences introduced). In doing so, suppose we have three categorical variables, X_1, X_2, and X_3, that are cross-classified to form a three-way contingency table consisting of I rows, J columns, and K tubes, respectively, and where the sample size is f_t.

Let $\Re^{I \times J \times K}$ be the space subscripted by i, j, and k, for $i = 1, 2, \ldots, I$, $j = 1, 2, \ldots, J$, and $k = 1, 2, \ldots, K$. Define $\mathbf{F}$ to be the three-way contingency table of size $I \times J \times K$ where the (i, j, k)th element is f_{ijk}. We also denote $\mathbf{P}$ to be the $I \times J \times K$ matrix consisting of the joint relative frequencies where the (i, j, k)th element is $p_{ijk} = f_{ijk}/f_t$ such that $\sum_{i=1}^{I} \sum_{j=1}^{J} \sum_{k=1}^{K} p_{ijk} = 1$. Define the ith row, jth column, and kth marginal proportions by

$$p_{i\bullet\bullet} = \sum_{j=1}^{J} \sum_{k=1}^{K} p_{ijk} \quad p_{\bullet j\bullet} = \sum_{i=1}^{I} \sum_{k=1}^{K} p_{ijk} \quad p_{\bullet\bullet k} = \sum_{i=1}^{I} \sum_{j=1}^{J} p_{ijk} \, .$$

We also denote the bivariate marginal proportions by

$$p_{ij\bullet} = \sum_{k=1}^{K} p_{ijk} \quad p_{i\bullet k} = \sum_{j=1}^{J} p_{ijk} \quad p_{\bullet jk} = \sum_{i=1}^{I} p_{ijk} \, .$$

Lastly, we define $\mathbf{I}$ to be the identity matrix of size $I \times I$ in the row space $\Re^{I}$, and $\mathbf{D}_I$, $\mathbf{D}_J$, $\mathbf{D}_K$ to be the diagonal matrices of the marginal proportions $p_{i\bullet\bullet}$, $p_{\bullet j\bullet}$ and $p_{\bullet\bullet k}$ in the row, column, and tube space $\Re^{I}$, $\Re^{J}$, and $\Re^{K}$, respectively.

Recall from Chap. 7 that, for the purposes of analyzing the association between two categorical variables, we showed that one can obtain row scores and column scores through reciprocal averaging and that the solution obtained from this method is equivalent to the SVD of the matrix of Pearson residuals; see Eq. (7.26). When analyzing the association between three nominal categorical variables, scores for the row, column, and tube categories can be determined using the Tucker3 decomposition. Such a method of decomposition was described by Tucker (1963, 1964) and used for the purposes of performing multi-way correspondence analysis by Carlier and Kroonenberg (1996) and Kroonenberg (2008). The key to such a variant of correspondence analysis is to define the three-way equivalent of Pearson's residuals defined by Eq. (7.21) which is

$$\pi_{ijk} = \frac{p_{ijk} - p_{i\bullet\bullet}p_{\bullet j\bullet}p_{\bullet\bullet k}}{\sqrt{p_{i\bullet}p_{\bullet j}p_{\bullet\bullet k}}}$$

and is the (i, j, k)th element of the $I \times J \times K$ matrix Π. Carlier and Kroonenberg (1996) considered these elements for the purposes of multi-way correspondence analysis as did Beh and Lombardo (2014, 2020), and Lombardo et al. (2020a), who working with ordered variables applied a *hybrid* Tucker3 decomposition (which implies the use of orthogonal polynomials) to this three-way matrix.

When used to determine the set of scores for the row, column, and tube categories, the Tucker3 decomposition of Π does not give a *closed form solution*. Instead, the solution to these scores (and the correlation between them which we will describe shortly) can be determined using an alternating least squares (ALS) procedure based on computing optimal scores for each of the three categorical variables. The three-way table is still flattened so that it becomes a two-way matrix, but we do so for each way (variable), thereby obtaining three matrices of components (one for each variable)—the term "component" is most commonly used to describe the set of scores calculated for a variable and so we shall adopt term "component" throughout the remainder of this chapter. For three categorical variables, the starting solutions to the ALS procedure are found by performing a SVD on the three flattened matrices

$$\Pi_1 = \Pi_{I\times J\cdot K} \quad \Pi_2 = \Pi_{J\times I\cdot K} \quad \text{and} \quad \Pi_3 = \Pi_{K\times I\cdot J}\,.$$

Here Π_1 is a two-way matrix formed by combining the information in the column and tube variables, Π_2 is the two-way matrix formed from combining the row and tube variables and Π_3 is formed from combining the row and column information. The row, column, and tube components are contained in the columns of the matrices $\tilde{\mathbf{A}}$, $\tilde{\mathbf{B}}$, and $\tilde{\mathbf{C}}$, which are of size $I \times P$, $J \times Q$ and $K \times R$, respectively, where $P \leq I$, $Q \leq J$, and $R \leq K$. These matrices have the property $\mathbf{A}^T\mathbf{D}_I\mathbf{A} = \mathbf{I}$, $\mathbf{B}^T\mathbf{D}_J\mathbf{B} = \mathbf{I}$ and $\mathbf{C}^T\mathbf{D}_K\mathbf{C} = \mathbf{I}$.

The starting components in these matrices are then used to determine the three-way equivalent of singular values, called *core elements*. For example, by considering Π_1, these core elements can be calculated from

$$\Lambda_{P\times Q\cdot R} = \mathbf{A}^T\Pi_1\left(\mathbf{B}^T \otimes \mathbf{C}^T\right)^T$$

so that one can recalculate Π_1 by

$$\hat{\Pi}_1 = \mathbf{A}\Lambda\left(\mathbf{B} \otimes \mathbf{C}\right)^T$$

which is the decomposed using SVD to yield an updated solution to $\mathbf{A}$. The matrix Λ is a flattened *core array* and is interpreted as the generalized form (for the three-way case) of the diagonal matrix of singular vectors, Λ, described in Eq. (10.2). However, Λ here is not diagonal since the components are not bi-orthogonal (just as the orthogonal polynomials of the BMD are not bi-orthogonal) and its elements represent the link between the row, column, and tube components, where its values reflect the strength of the association between the three variables.

The optimal components (the set of *scores*) contained in the columns of **A**, **B**, and **C** are then computed by using alternating least squares so that

$$||\Pi - \hat{\Pi}||^2$$

is minimized. The approximation of Π, $\hat{\Pi}$, is determined from the choice of the number of components extracted from the matrix and so the convergence of the procedure depends on the value of P, Q, and R chosen.

The Tucker3 decomposition of the three-way matrix, Π can be more succinctly expressed by noting that the (i, j, k)th element can be expressed using summation notation by

$$\pi_{ijk} = \sum_{p=1}^{P} \sum_{q=1}^{Q} \sum_{r=1}^{R} \lambda_{pqr} a_{ip} b_{jq} c_{kr} + e_{ijk} .$$

Here, a_{ip} is the score of the ith row along the pth component, b_{jq} is the score given to the jth column along the qth component and c_{kr} is the score of the kth component along the rth component, while $P \leq I$, $Q \leq J$, and $R \leq K$ are the fixed number of the components in the matrices **A**, **B**, and **C**, respectively. The core element λ_{pqr} and equivalent to the *generalized correlation* between a_{ip}, b_{jq}, and c_{kr} such that

$$\lambda_{pqr} = \sum_{i=1}^{I} \sum_{j=1}^{J} \sum_{k=1}^{K} p_{ijk} a_{ip} b_{jq} c_{kr} .$$

A very useful property of these values is that their sum-of-squares gives the total inertia such that

$$\frac{X^2}{f_t} = \sum_{p=1}^{I} \sum_{q=1}^{J} \sum_{r=1}^{K} \lambda_{pqr}^2 . \tag{10.15}$$

Finally, e_{ijk} is the error term of the approximation. When $P = I$, $Q = J$, and $R = K$, all of the error terms are zero so that $\Pi = \hat{\Pi}$. However, this situation will rarely, if ever, arise in practice since the need for higher-order (fourth, fifth, etc.) components is seldom required and provides a minimal contribution to the reconstitution to the elements of Π (van Herk and van de Velden 2007).

One of the helpful features of MWCA is that the total inertia can be calculated using the sum-of-squares of the elements of the core array that is of full-rank so that

$$\frac{X^2}{f_t} = \sum_{p=1}^{I} \sum_{q=1}^{J} \sum_{r=1}^{K} \lambda_{pqr}^2 .$$

Although, additional insight can be gained about the structure of the association by partitioning the chi-squared so that

$$X^2 = X^2_{IJ} + X^2_{IK} + X^2_{JK} + X^2_{IJK} \,. \tag{10.16}$$

The first three terms of Eq. (10.16) are the pairwise chi-squared statistics obtained by aggregating across the categories of each variable. That is, the pairwise association between the row and column variables is quantified by the term X^2_{IJ} under the null hypothesis of complete independence, and is a χ^2 random variable with $(I-1)(J-1)$ degrees of freedom. Similarly, the pairwise row-tube term X^2_{IK} is asymptotically $\chi^2_{(I-1)(K-1)}$ and the column-tube association is assessed by the term X^2_{JK} which is a $\chi^2_{(J-1)(K-1)}$. The last term, X^2_{IJK}, measures the trivariate association and is $\chi^2_{(I-1)(J-1)(K-1)}$. Further insight into the partition of Pearson's three-way chi-squared statistic for MWCA can be found by referring to Carlier and Kroonenberg (1996), Kroonenberg (2008) [p. 435], and Beh and Lombardo (2014)[p. 458]. Other details of this partition for ordinal/nominal categorical variables can be found by referring to Beh and Davy (1998), Beh and Davy (1999), Loisel and Takane (2016), and Lombardo et al. (2020c).

Our discussion here implies that there exists a symmetric association between the three categorical variables, although one may also adopt a similar strategy outlined here when there exists an asymmetric association between the variables. For discussions of this variant see Lombardo et al. (1996), Lombardo et al. (2013, 2018), and Lombardo et al. (2020a).

10.6 The Interactive Biplot

To portray the three-way symmetric (and asymmetric) association between three categorical variables, we consider an adaptation of the isometric biplot that has been described earlier in this chapter. Such a biplot is referred to as an *interactive* biplot (Carlier and Kroonenberg 1996; Kroonenberg 2008; Lombardo et al. 1996; Beh and Lombardo 2014)—but has also been called a *nested-mode biplot* (Kroonenberg 2008, Sect. 11.5.4)—and allows for the association between all of the variables to be displayed in a single plot. The use of the term "interactive" means that this biplot is constructed by taking into consideration that there exists an association between two of the variables so that the interaction between the categories of these two variables is captured in the biplot.

There are two types of interactive biplot that we shall describe. The first is the *column-tube* interactive biplot, where it is assumed, or is known, that there exists an association between the column and tube variables; certainly such an association can be assessed from the partition of Pearson's chi-squared statistic; see Eq. (10.16). For such a biplot, the association between these two variables is depicted as a single point by their principal coordinates and their interaction with the row categories can be determined by examining the distance of these points from the projection made of the row standard coordinates from the origin. The second type of biplot that we shall describe is the *row* interactive biplot. For this display, the interaction between the column and tube categories is still considered but their position in the biplot is depicted as a projection from the origin to their standard coordinates, while the row categories are depicted as points using their principal coordinates. In both plots, there will be I row points and JK points that reflect the interaction between the J column categories and K tube categories.

Of course, one may wish to construct a column, or a row-tube, interactive biplot or even a tube, or row-column, isometric biplot by following a similar strategy. We now turn our attention to defining the standard and principal coordinates for the row and column-tube interactive biplots, but first remind the interested reader that excellent descriptions of the biplot have been given by, for example, Gabriel (1971), Greenacre (2007a), Greenacre (2017), Gower et al. (2011), Gower et al. (2015, 2016).

To construct a column-tube interactive biplot, the standard coordinate of the ith row and the principal coordinate of the (j, k)th pair of column-tube principal coordinates along the pth dimension is defined as

$$\begin{aligned} f_{ip} &= a_{ip} \\ g_{jkp} &= \sum_{q=1}^{Q} \sum_{r=1}^{R} \lambda_{pqr} b_{jq} c_{kr} \end{aligned}$$

for $p = 1, 2, \ldots, P$. Therefore, a column-tube interactive biplot is comprised of at most P dimensions.

If a row interactive biplot is to be constructed, then the principal coordinate of the ith row and the standard coordinate of the (j, k)th pair of column-tube principal coordinates along the (q, r)th dimension is

$$\begin{aligned} f_{iqr} &= \sum_{p=1}^{P} \lambda_{pqr} a_{ip} \\ g_{jkqr} &= b_{jq} c_{kr} \end{aligned}$$

for $q = 1, 2, \ldots, Q$ and $r = 1, 2, \ldots, R$. Therefore, a row interactive biplot is a visualization of the association between the three variables that consists of at most QR dimensions. Since QR will often be larger than P, the two-dimnsional (say) column-tube interactive biplot will often be a better quality than a two-dimensional row interactive biplot. Interpreting the interaction between specific categories from one variable with specific categories from the pairing of the remaining two categories can also be made by considering the inner product of the standard coordinates and the principal coordinates. These inner products provide numerical confirmation of the interactions identified in the interactive biplot and such confirmations are demonstrated by Beh and Lombardo (2020).

10.6.1 The Biplot and Three-Way Correspondence Analysis

10.6.1.1 The Tax Courts System in Italy

To illustrate the construction and interpretation of an interactive biplot, we shall apply a three-way correspondence analysis to data that was obtained from a 2018 quarterly report of the Italian Ministry of Economy and Finance. It provides details of the evaluation of the tax courts system in Italy, and was also examined by Lombardo et al. (2020b), who applied to it a three-way version of log-ratio analysis. This data is summarized in the $4 \times 5 \times 2$ contingency table of Table 10.4 and summarizes the number of tax disputes according to the *Geographical Area* of Italy (North, *N*, Centre, *C*, South, *S*, and Isles, *Is*), the type of *Tax Revenue Service* (that consists of the internal revenue agency (*IRA*), the revenue collection agency (*RCA*), the monopolies customs agency (*ITA*), local collection services (*LTA*) and other services (*OTA*)) and the level of *Justice Court* (being the provincial jurisdiction (*CTP*) and the regional court of appeal (*CTR*)).

In the tax courts system, all sections have to either accept or reject the tax appeal on the spot, so hearings cannot be postponed. As a consequence of this, there is no delay in processing cases. However, the lower the number of tax appeals that are presented, the more efficient the tax courts system is in processing them. Therefore, we aim to investigate the association between the geographical distribution of tax disputes across Italy and the efficiency of the system.

Table 10.4 Tax disputes by geographical area, type of revenue service, and level of justice court

	CTP				
	IRA	*RCA*	*ITA*	*LTA*	*OTA*
N	3145	1127	130	602	211
C	1912	2668	56	692	354
S	2539	2683	60	2225	1399
Is	1331	147	43	1292	2775
	CTR				
	IRA	*RCA*	*ITA*	*LTA*	*OTA*
N	1798	255	91	228	95
C	1621	605	64	354	190
S	1795	1018	42	506	533
Is	1138	20	28	279	460

10.6.2 Size and Nature of the Dependence

Our analysis starts with investigating the importance of the two-way and three-way association terms with the aim of determining if and how *Geographical Area* is associated with the *Tax Revenue Service* across the Italian *Justice Courts*. To do so, we partition Pearson's X^2 statistic into three two-way terms and one three-way term; see Eq. (10.16). Table 10.5 summarizes the results of this partition which include the four chi-squared terms, the percentage contribution that each of them makes to X^2 and the degrees of freedom for each term. Since there are large number of observations in Table 10.5, the three bivariate chi-squared terms and the three-way term are all statistically significant at the 0.01 level of significance. Therefore, we will, in general, not refer to the statistical significance of each association term, but only compare the relative sizes of the inertia. Indeed, the mean of a χ^2 random variable is df, thus Table 10.5 also summarizes the X^2/df values for each term so that comparisons can be made between the X^2 values from different chi-squared distributions. The overall association is assessed by the chi-squared statistic of $X^2 = 11855$, with $df = 31$, and shows that there is evidence of a strong association between the three categorical variables of Table 10.4. It also shows that, with X^2_{IJ} accounting for 75% of the association that exists in Table 10.4, the most dominant source is due to the association between the four *Geographical Areas* and the *Tax Revenue Service*, suggesting that the different services behave quite differently across the four regions of Italy. While being statistically significant, the association between *Tax Revenue Service* and *Court Justice* only accounts for 1% of the association that exists between the three categorical variables, and so plays a negligible role in determining the magnitude of X^2. Note that for the X^2 terms in Table 10.4, the subscripts I, J, and K refer to the variables *Geographical Area*, *Tax Revenue Service*, and *Justice Court*, respectively.

While partitioning Pearson's X^2 provides an excellent overview of those variables that are associated, and their dominance in defining the overall association structure, it does not provide any insight into how particular categories from one variable interact with categories from another variable. We now turn our attention to understanding this facet of the association by performing a three-way correspondence analysis on Table 10.4, and in doing so, construct a column-tube interactive biplot to visualize this association.

Table 10.5 Partitioning Pearson's chi-squared statistic of Table 10.4

Association	Index	% Inertia	*df*	X^2/df
X^2_{IJ}	8949	75	12	746
X^2_{IK}	125	1	3	42
X^2_{JK}	1648	14	4	412
X^2_{IJK}	1133	10	12	94
Total	11855	100	31	382

10.6.3 The Interactive Biplot

We perform a three-way correspondence analysis on the data summarized in Table 10.4 by applying a Tucker3 decomposition to it matrix Π; the analysis is carried out using the R package `CA3variants`. In doing so, the analysis is performed with $P = 3$, $P = 2$ and $P = 2$. This choice was selected since it provides a better two-dimensional representation of the association between the three variables than the choice of $P = I = 4$, $P = J = 5$, and $R = K = 2$. With this choice of P, Q, and R the amount of inertia that is explained from the Tucker3 decomposition is

$$f_t \sum_{p=1}^{3} \sum_{q=1}^{2} \sum_{p=1}^{2} = 11373.42$$

and so reflects $100 \times 11373.42/11855 = 95.9\%$ of the total inertia.

By selecting $P = 3$, $P = 2$, and $P = 2$, and calculating the corresponding row, column, and tube components, the resulting column-tube interactive biplot is given by Fig. 10.4 and shows the association between the variables *Geographical Area*, *Tax Revenue Service*, and *Justice Court*. It has been constructed with the knowledge that there exists a statistically significant association between the column variable (*Tax Revenue Service*) and the tube variable (*Justice Court*). However, in an efficient tax court system, there should not be an association between *Geographical Area* and the *Tax Revenue Service* irrespective of the type of justice court. Therefore, gaining an understanding of how differently tax disputes are managed in the different

Fig. 10.4 The column-tube biplot from the three-way correspondence analysis of Table 10.4

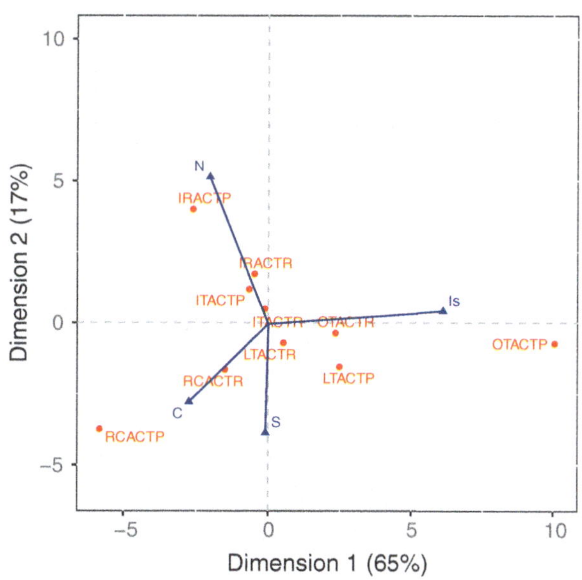

geographical areas of Italy might help policy makers better organize the tax court system. Figure 10.4 being a column-tube interactive biplot, it jointly displays the four row categories as a projection from the origin to their standard coordinates, and the 10 column-tube categories are depicted by their (joint) principal coordinates.

Suppose we focus on the distance between the interaction of the column and tube categories in the interactive biplot. Since these have been depicted as a single point defined using their principal coordinates, the distance between these points is interpretable and quantifiable; recall that the squared distances between two principal coordinates is Euclidean. When we do so, we observe that there is a large distance between *IRACTP* (internal revenue agency—regional court of appeal, that lies at the centre of the top-left quadrant) and *OTACTP* (other services–regional court of appeal, which lies on the far right of the bottom right quadrant). This suggests that the internal revenue agency and other services not listed are assessed differently at the regional court of appeal. However, since *ITACTP* and *ITACTR* lie in close proximity to one another, this suggests that the monopolies customs agency are assessed in a similar manner at the two justice courts. One can also see how the interaction between categories of the *Tax Revenue Service* and the *Justice Court* are assessed across the four *Geographical Areas* of Italy. In particular, Fig. 10.4 shows that the north (*N*) of Italy is characterized by disputes managed by the *monopolies customs agency* (*ITA*) and the *internal revenue agency* (*IRA*). On the other hand, the central (*C*) region of Italy has a strong association with the revenue collection agency (*RCA*) while for the south (*S*) is the *local collection services* (*LTA*). Such information may well prove helpful for policy makers to formulate and adopt directives concerning the re-organization of the tax revenue services in Italy to ensure a more homogeneous judicial system.

Concerning other tax services (*OTA*), tax disputes that are managed by the provincial tax court (*CTP*) are far more likely to be observed in the isles (*Is*) than in any other region of Italy. Disputes managed by the regional court of appeal (*CTR*) in the north (*N*), centre (*C*), and the isles (*Is*) of Italy vary strongly across the various tax services (*ITA*, *IRA*, *RCA*, *LRA*, and *OTA*), suggesting they are managed differently by these different services throughout Italy. Furthermore, tax disputes managed by the internal revenue agency of the regional court (*IRACTR*) in central (*C*) and southern (*S*) Italy are strongly associated with the disputes managed by the revenue collection agency (*RCA*) on the isles (*Is*).

Concerning the overall performance of the two courts, (*CTP* and *CTR*), in handling disputes the number of disputes managed by the provincial tax court (*CTP*) is always higher than the number of disputes managed by the regional tax court (*CTR*). This suggests that there is an efficiency in the system since many cases are being managed at the provincial level (*CTP*) rather than being elevated to the regional level (*CTR*). Despite this, as we have previously discussed, differences still exist at among the four geographical areas.

One aspect that we have yet to address is the quality of the visual display of Fig. 10.4. Since this figure is a column-tube isometric biplot, the percentage contribution of its two dimensions to the total inertia can be quantified by

$$100 \times \frac{\sum_{q=1}^{J}\sum_{r=1}^{K}\lambda_{1qr}^{2}}{X^2/f_t} \quad \text{and} \quad 100 \times \frac{\sum_{q=1}^{J}\sum_{r=1}^{K}\lambda_{2qr}^{2}}{X^2/f_t},$$

respectively. This is because the total inertia can be expressed by Eq. (10.15). Calculating these quantities for Table 10.4 shows that the first dimension of the biplot visually shows 65% of the association between the three variables while the second dimension shows 17% of this association. Therefore, Fig. 10.4 visually describes about 82% of association that exists between the three categorical variables and is thus a very good visual display.

10.7 Final Remarks

We have demonstrated in Chap. 7 that there is a clear link between the reciprocal averaging of the centred row and column profiles and SVD for two nominal categorical variables. Such a link for ordinal variables is more straight forward and does not require an iterative process to determine the scores. This is because the orthogonal polynomials can be easily calculated using the recurrence formulae of Emerson (1968). However, if one were to approach the scoring problem from a reciprocal averaging perspective, then the two formulae take the form

$$\mathring{\lambda}_{uv}\mathring{x}_{iu} = \left(\frac{p_{i1}}{p_{i\bullet}} - p_{\bullet 1}\right)\mathring{y}_{1v} + \ldots + \left(\frac{p_{in}}{p_{i\bullet}} - p_{\bullet n}\right)\mathring{y}_{nv} = \sum_{j=1}^{n}\left(\frac{p_{ij}}{p_{i\bullet}} - p_{\bullet j}\right)\mathring{y}_{jv}$$

and

$$\mathring{\lambda}_{uv}\mathring{y}_{jv} = \left(\frac{p_{1j}}{p_{\bullet j}} - p_{1\bullet}\right)\mathring{x}_{1u} + \ldots + \left(\frac{p_{mj}}{p_{\bullet j}} - p_{m\bullet}\right)\mathring{x}_{mu} = \sum_{i=1}^{m}\left(\frac{p_{ij}}{p_{\bullet j}} - p_{i\bullet}\right)\mathring{x}_{iu}$$

where

$$\mathrm{E}\left(\mathring{x}_{iu}\right) = \sum_{i=1}^{m} p_{i\bullet}\mathring{x}_{iu} = 0 \qquad \mathrm{Var}\left(\mathring{x}_{iu}\right) = \sum_{i=1}^{m} p_{i\bullet}\mathring{x}_{iu}^{2} = 1$$

$$\mathrm{E}\left(\mathring{y}_{jv}\right) = \sum_{j=1}^{n} p_{\bullet j}\mathring{y}_{jv} = 0 \qquad \mathrm{Var}\left(\mathring{y}_{jv}\right) = \sum_{j=1}^{n} p_{\bullet j}\mathring{y}_{jv}^{2} = 1\,.$$

In matrix form, these two properties are $\mathring{\mathbf{X}}^T\mathbf{D}_r\mathring{\mathbf{X}} = \mathbf{I}$ and $\mathring{\mathbf{Y}}^T\mathbf{D}_c\mathring{\mathbf{Y}} = \mathbf{I}$, which are defined for Eq. (10.6).

For the study of multiple categorical variables, there exists a link of Gram-Schmidt orthonornalization to the Tucker3 decomposition (Kroonenberg et al. 1989; Kroonenberg 2008, p. 100). Further developments could highlight the global link of the reciprocal averaging method of scaling categorical (nominal and ordinal) variables to the various methods that are available for multivariate decomposition; see also Lombardo et al. (2020a).

References

Agresti, A. (2007). *An Introduction to Categorical Data Analysis*. New York: Wiley.

Agresti, A. (2013). *Categorical Data Analysis* (3rd ed.). New York: Wiley.

Beh, E. J. (1997). Simple correspondence analysis of ordinal cross-classifications using orthogonal polynomials. *Biometrical Journal, 39,* 589–613.

Beh, E. J. (1998). A comparative study of scores for correspondence analysis with ordered categories. *Biometrical Journal, 40,* 413–429.

Beh, E. J. (2001). Partitioning Pearson's chi-squared statistic for singly ordered two-way contingency tables. *The Australian and New Zealand Journal of Statistics, 43,* 327–333.

Beh, E. J. (2004). Simple correspondence analysis: A bibliographic review. *International Statistical Review, 72,* 257–284.

Beh, E. J., & Davy, P. J. (1998). Partitioning Pearson's chi-squared statistic for a completely ordered three-way contingency table. *The Australian and New Zealand Journal of Statistics, 40,* 465–477.

Beh, E. J., & Davy, P. J. (1999). Partitioning Pearson's chi-squared statistic for a partially ordered three-way contingency table. *The Australian and New Zealand Journal of Statistics, 41,* 233–246.

Beh, E. J., & Lombardo, R. (2012). A genealogy of correspondence analysis. *Australian and New Zealand Journal of Statistics, 54,* 137–168.

Beh, E. J., & Lombardo, R. (2014). *Correspondence Analysis: Theory*. Wiley, Chichester: Practice and New Strategies.

Beh, E. J., & Lombardo, R. (2020). Five strategies for accommodating overdispersion in simple correspondence analysis. In T. Imaizumi, A. Okado, S. Miyamoto, F. Sakaori, Y. Yamamoto, & M. Vichi (Eds.), *Advanced Studies in Classification and Data Science, page 117–129*. Singapore: Springer.

Beh, E. J., & Lombardo, R. (2019). Multiple and multiway correspondence analysis. *WIRE's Computational Statistics*, 11(5):e1464 (11 pages).

Beh, E. J., & Lombardo, R. (2020). *An Introduction to Correspondence Analysis*. Chichester (in press): Wiley.

Best, D. J., & Rayner, J. C. W. (1996). Nonparametric analysis for doubly ordered two-way contingency tables. *Biometrics, 52,* 1153–1156.

Carlier, A., & Kroonenberg, P. M. (1996). Decomposition and biplots in three-way correspondence analysis. *Psychometrika, 61,* 355–373.

Carroll, J., Green, P. E., & Schaffer, C. M. (1986). Interpoint distance comparisons in correspondence analysis. *Journal of Marketing Research, 23,* 271–280.

Carroll, J., Green, P. E., & Schaffer, C. M. (1987). Interpoint distance comparisons in correspondence analysis: A clarification. *Journal of Marketing Research, 24,* 445–450.

Carroll, J., Green, P. E., & Schaffer, C. M. (1989). Reply to greenacre's commentary on the carroll-green-schaffer scaling of two-way correspondence analysis solutions. *Journal of Marketing Research, 26,* 366–368.

Carroll, J. D., & Chang, J. J. (1970). Analysis of individual differences in multi-dimensional scaling via an n-way generalisation of "Eckart - Young" decomposition. *Psychometrika, 35,* 283–319.

D'Ambra, L., & Lauro, N. C. (1989). Non-symmetrical correspondence analysis for three-way contingency table. In R. Coppi & S. Bolasco (Eds.), *Multiway Data Analysis* (pp. 301–315). Amsterdam: Elsevier.

Emerson, P. L. (1968). Numerical construction of orthogonal polynomials from a general recurrence formula. *Biometrics, 24,* 695–701.

Gabriel, K. R. (1971). The biplot graphic display of matrices with application to principal component analysis. *Biometrika, 58,* 453–467.

Goodman, L. A., & Kruskal, W. (1954). Measures of association for cross-classifications. *Journal of the American Statistical Association, 49,* 732–764.

Gower, J. C., Le Roux, N., & Lubbe, S. (2015). Biplots: quantitative data. *WIREs Computational Statistics, 7,* 42–62.

Gower, J. C., Le Roux, N., & Lubbe, S. (2016). Biplots: qualititative data. *WIREs Computational Statistics, 8,* 82–111.

Gower, J. C., Lubbe, S., & Le Roux, N. (2011). *Understanding Biplots*. Wiley.

Gray, L., & Williams, J. (1981). Goodman and Kruskal τ_b multiple and partial analogs. *Proceedings of the Social Statistics Section, 10,* 50–62.

Greenacre, M. J. (1984). *Theory and Application of Correspondence Analysis*. London: Academic Press.

Greenacre, M. J. (2007a). *Correspondence Analysis in Practice* (2nd ed.). Boca Raton, FL: Chapman & Hall/CRC.

Greenacre, M. J. (2007b). *Correspondence Analysis in Practice* (3rd ed.). Boca Raton, FL: Chapman & Hall/CRC.

Greenacre, M. J. (2017). *Biplots in Practice* (3rd ed.). Barcelona: Fundacion BBVA.

Harshman, R. A. (1970). Foundations of the PARAFAC procedure: Models and conditions for an "explanatory" multimodal factor analysis. *UCLA Working Papers in Phonetics*, 16:1 – 84.

Kiers, H. A. L. (2000). Towards a standardized notation and terminology in multiway analysis. *Chemometrics and Intelligent Laboratory Systems, 14,* 105–122.

Kim, S.-K., & Grochowalski, J. H. (2019). Exploratory visual inspection of category associations and correlation estimation in multidimensional subspaces. *Journal of Classification, 36,* 177–199.

Konig, R. (2010). Changing social categories in a changing society: studying trends with correspondence analysis. *Quality & Quantity, 44,* 409–425.

Kroonenberg, P. M., & Lombardo, R. (1999). Nonsymmetric correspondence analysis: A tool for analysing contingency tables with a dependence structure. *Multivariate Behavioral Research, 34,* 367–396.

Kroonenberg, P. M. (2008). *Applied Multiway Data Analysis*. Wiley.

Kroonenberg, P. M., & de Leeuw, J. (1980). Principal components analysis of three-mode data by means of alternating least squares algorithms. *Psychometrika, 45,* 69–97.

Kroonenberg, P. M., ten Berge, J. M. F., Brouwer, P., & Kiers, H. (1989). Gram-schmidt versus bauer-rutishauser in alternating least-squares algorithms for three-mode principal component analysis. *Computational Statistics Quarterly, 2,* 81–87.

Lancaster, H. O. (1953). A reconciliation of χ^2, considered from metrical and enumerative aspects. *Sankhya, 13,* 1–10.

Lauro, N. C., & D'Ambra, L. (1984). L'analyse non symmétrique des correspondances. In Diday, E., editor, *Data Analysis and Informatics, III*, pages 433 – 446. North-Holland.

Lebart, L., Morineau, A., & Warwick, K. M. (1984). *Multivariate Descriptive Statistical Analysis*. Wiley.

Loisel, S., & Takane, Y. (2016). Partitions of pearson's chi-square statistic for frequency tables: A comprehensive account. *Computational Statistics, 31,* 1429–1452.

Lombardo, R. (2011). Three-way association measure decompositions: the Delta index. *Journal of Statistical Planning and Inference, 141,* 1789–1799.

Lombardo, R., & Beh, E. J. (2010). Simple and multiple correspondence analysis using orthogonal polynomials. *Journal of Applied Statistics, 37,* 2101–2116.

Lombardo, R., & Beh, E. J. (2016). Variants of simple correspondence analysis. *The R Journal, 8*(2), 167–184.
Lombardo, R., Beh, E. J., & D'Ambra, L. (2007). Non-symmetric correspondence analysis with ordinal variables using orthogonal polynomials. *Computational Statistics and Data Analysis, 52,* 566–577.
Lombardo, R., Beh, E. J., & Guerrero, L. (2013). Non-symmetric three-way correspondence analysis to analyse text data in a food context. In E. Bretani & M. Carpita (Eds.), *Advances on Latent Variables - Methods, Models and Applications, page 6 pages*. Italy: Vita e Pensiero Publishers, Milan.
Lombardo, R., Beh, E. J., & Guerrero, L. (2018). Analysis of three-way asymmetric association of food concepts in cross-cultural marketing. *Quality & Quantity, 53,* 2323–2337.
Lombardo, R., Beh, E. J., & Kroonenberg, P. M. (2016). Modelling trends in ordered correspondence analysis using orthogonal polynomials. *Psychometrika, 81,* 325–349.
Lombardo, R., Beh, E. J., & Kroonenberg, P. M. (2020a). Symmetrical and non-symmetrical variants of three-way correspondence analysis for ordered variables. *Statistical Science*, page (in press).
Lombardo, R., Camminatiello, I., D'Ambra, A., & Beh, E. J. (2020b). Assessing the Italian tax courts system by weighted three-way log-ratio analysis. *Socio-Economic Planning Sciences*, page (in press).
Lombardo, R., Carlier, A., & D'Ambra, L. (1996). Nonsymmetric correspondence analysis for three-way contingency tables. *Methodologica, 4,* 59–80.
Lombardo, R., & Meulman, J. (2010). Multiple correspondence analysis via polynomial transformations of ordered categorical variables. *Journal of Classification, 27,* 191–216.
Lombardo, R., Takane, Y., & Beh, E. J. (2020c). Familywise decompositions of Pearson's chi-squared statistic in the analysis of contingency tables. *Advances in Data Analysis and Classification, 14,* 629–649.
Marcotorchino, F. (1985). Utilisation des comparaisons par paires en statistique des contingencies: Partie III. page 39 pages. Report # F - 081, Etude du Centre Scientifique, IBM, France.
Nishisato, S. (1980). *Analysis of Categorical Data: Dual Scaling and its Applications*. Toronto: University of Toronto Press.
Nishisato, S. (2016). Multidimensional joint graphical display of symmetric analysis: Back to the fundamentals. In van der Ark, L., Bolt, D. M., Wen-Chung, W., Douglas, J. A., and Wiberg, M., editors, *Quantitative Psychology Research*, number 167 in Proceedings in Mathematics & Statistics, pages 291 – 298. Psychometric Society, Springer-Verlag. Heidelberg.
Nishisato, S., & Arri, P. S. (1975). Nonlinear programming approach to optimal scaling of partially ordered categories. *Psychometrika, 40,* 525–548.
Nishisato, S., & Clavel, J. G. (2003). A note on between-set distances in dual scaling and correspondence analysis. *Behaviormetrika*, 87–98.
Pearson, K. (1904). On the theory of contingency and its relation to association and normal correlation. *Drapers Memoirs*, Biometric Series, Vol 1, London.
Rayner, J. C. W., & Best, D. J. (1996). Smooth extensions of pearson's product moment correlation and Spearman's rho. *Statistics & Probability Letters, 30,* 171–177.
Rayner, J. C. W., & Best, D. J. (2001). *A Contingency Table Approach to Nonparametric Testing*. Chapman & Hall/CRC.
Simonetti, B., Beh, E. J., & D'Ambra, L. (2011). The analysis of dependence for three way contingency tables with ordinal variables: a case study of patient satisfaction. *Journal of Applied Statistics, 37,* 91–103.
Tucker, L. R. (1963). Implications of factor analysis of three-way matrices for measurement of change. In C. W. Harris (Ed.), *Problems in Measuring Change* (pp. 122–137). Madison, Wisconsin: The University of Wisconsin Press.
Tucker, L. R. (1964). The extension of factor analysis to three-dimensional matrices. In Frederiksen, N. and Gulliksen, H., editors, *Contributions to Mathematical Psychology*, pages 109 – 127. Holt,Rinehart & Winston, Canada.

Tucker, L. R. (1966). Some mathematical notes on three-mode factor analysis. *Psychometrika, 31,* 279–311.

van Herk, H., & van de Velden, M. (2007). Insight into the relative merits of rating and ranking in a cross-national context using three-way correspondence analysis. *Food Quality and Preference, 18,* 1096–1105.

Weller, S. C., & Romney, A. K. (1990). *Metric Scaling: Correspondence Analysis*. Sage University Paper Series on Quantitative Applications in the Social Sciences 07-075. Newbury Park, CA: Sage.

Chapter 11
On the Analysis of Over-Dispersed Categorical Data

11.1 Introduction

In Chaps. 7 and 8, our focus has been on describing some of the technical aspects of reciprocal averaging (and canonical correlation analysis), so that one can obtain row and column scores that maximize the association between the variables of a two-way contingency table. The foundations under which, such discussions are laid, rests upon the assumption that the cell frequencies of the contingency table are assumed to be Poisson random variables. This is apparent by observing that for the (i, j)th cell it is defined Z_{ij}.

$$Z_{ij} = \frac{n_{ij} - f_t p_{i\bullet} p_{\bullet j}}{\sqrt{f_t p_{i\bullet} p_{\bullet j}}} = \sqrt{f_t} \frac{p_{ij} - p_{i\bullet} p_{\bullet j}}{\sqrt{p_{i\bullet} p_{\bullet j}}} \tag{11.1}$$

which is referred to as the Pearson residual for that cell. Therefore, by virtue of the Poisson distribution, this structure implies that $\mathrm{E}\left(n_{ij}\right) = \mathrm{Var}\left(n_{ij}\right) = f_t p_{i\bullet} p_{\bullet j}$, for all $i = 1, 2, \ldots, m$ and $j = 1, 2, \ldots, n$.

A problem with this strategy is that the equality of the expectation and variance rarely holds true. In fact, Agresti (2013) [p. 80] shows that the asymptotic variance of Z_{ij} "is less than 1.0, averaging $[(m-1)(n-1)]$ / (number of cells)". Haberman (1973) showed that the asymptotic variance of Z_{ij} is $(1 - p_{i\bullet})\left(1 - p_{\bullet j}\right)$ so that

$$Z_{ij}^* = \sqrt{f_t} \frac{p_{ij} - p_{i\bullet} p_{\bullet j}}{\sqrt{p_{i\bullet} p_{\bullet j} (1 - p_{i\bullet}) \left(1 - p_{\bullet j}\right)}}$$

is asymptotically normally distributed. The discussions of Haberman (1973) and Agresti (2013) imply that the variance of Z_{ij} exceeds its variance, thereby highlighting that, generally speaking, a two-way contingency table is prone to over-dispersion. A simple visual aid for detecting over-dispersion was presented by Efron (1992), where he suggested plotting Z_{ij} against $\sqrt{\mathrm{E}\left(n_{ij}\right)} = \sqrt{f_t p_{i\bullet} p_{\bullet j}}$. If such a plot gives

S. Nishisato et al., *Modern Quantification Theory*,
Behaviormetrics: Quantitative Approaches to Human Behavior 8,
https://doi.org/10.1007/978-981-16-2470-4_11

a scatter of points with no obvious pattern or consistent spread, then there is no apparent dispersion (either over-dispersion, or under-dispersion) in the data.

11.2 Generalized Pearson Residual

The presence of over-dispersion in a contingency table can certainly influence the ability to determine the most appropriate row and column scores. Over-dispersion also influences the way in which a visual summary of the association is constructed, even in the context of correspondence analysis; this issue was raised by Beh and Lombardo (2020a). They described the nature of this over-dispersion and provided various strategies in which it can be dealt with. Therefore, the purpose of this chapter is to continue with this discussion and demonstrate the applicability of these strategies. Rather than our starting point being in terms of comparing profiles, as was our focus in Chaps. 7 and 10, here we shall examine the over-dispersion issue by presenting a generalization of the Pearson residual, Z_{ij}, and applying a singular value decomposition to this generalized residual. The generalization of Z_{ij} shall be made by incorporating an over-dispersion parameter, γ such that the generalized Pearson residual can be expressed in the form

$$\dot{Z}_{ij}\left(\gamma | f_t\right) = \frac{n_{ij} - \mathrm{E}\left(n_{ij} | f_t; \gamma\right)}{\sqrt{\mathrm{Var}\left(n_{ij} | f_t; \gamma\right)}} \tag{11.2}$$

for $i = 1, 2, \ldots, m$ and $j = 1, 2, \ldots, n$, where the sample size, f_t, is fixed and known. One may therefore see that the classic Pearson residual $\dot{Z}_{ij} = Z_{ij}$—see Eq. (11.1)—is a special case Eq. (11.2) when

$$\mathrm{E}\left(n_{ij} | f_t; \gamma\right) = \mathrm{Var}\left(n_{ij} | f_t; \gamma\right) = f_t p_{i\bullet} p_{\bullet j} \, .$$

For the purposes of scaling, and performing correspondence analysis on a two-way contingency table, Eq. (11.2) can be expressed in terms of the joint proportions so that

$$\dot{Z}_{ij}\left(\gamma | f_t\right) = \sqrt{f_t}\frac{p_{ij} - \mathrm{E}\left(p_{ij} | f_t; \gamma\right)}{\sqrt{\mathrm{Var}\left(p_{ij} | f_t; \gamma\right)}} = \sqrt{f_t} Z_{ij}\left(\gamma | f_t\right)$$

where

$$\mathrm{E}\left(n_{ij} | f_t; \gamma\right) = f_t \mathrm{E}\left(p_{ij} | f_t; \gamma\right) \qquad \mathrm{Var}\left(n_{ij} | f_t; \gamma\right) = f_t^2 \mathrm{Var}\left(p_{ij} | f_t; \gamma\right) \, . \tag{11.3}$$

Therefore, a correspondence analysis can be performed by applying a singular value decomposition (SVD) to $Z_{ij}\left(\gamma | f_t\right)$ so that

$$Z_{ij}(\gamma|\,f_t) = \frac{p_{ij} - \mathrm{E}\left(p_{ij}|\,f_t;\gamma\right)}{\sqrt{\mathrm{Var}\left(p_{ij}|\,f_t;\gamma\right)}} = \sum_{k=1}^{K} \tilde{x}_{ik}(\gamma)\,\rho_k(\gamma)\,\tilde{y}_{jk}(\gamma)$$

for some γ. We define $\tilde{\mathbf{x}}_k = (\tilde{x}_{1k}, \tilde{x}_{2k}, \ldots, \tilde{x}_{mk})^T$ and $\tilde{\mathbf{y}}_k = (\tilde{y}_{1k}, \tilde{y}_{2k}, \ldots, \tilde{y}_{nk})^T$ to be the kth left and right singular vector of the matrix of $Z_{ij}(\gamma|\,f_t)$ values, respectively, for $k = 1, \ldots, K$. These vectors have the property $\tilde{\mathbf{x}}_k^T\tilde{\mathbf{x}}_k = 1$ and $\tilde{\mathbf{y}}_k^T\tilde{\mathbf{y}}_k = 1$. From this decomposition, a version of Pearson's chi-squared statistic of the contingency table can be expressed as

$$\tilde{X}^2(\gamma) = \sum_{i=1}^{m}\sum_{j=1}^{n} Z_{ij}^2(\gamma|f_t) = f_t\sum_{k=1}^{K}\rho_k(\gamma)^2\,. \tag{11.4}$$

It must be noted that some strategies used for dealing with over-dispersion involve an expression of the mean and variance that are not as simple being multiples of Eq. (11.3). Instead, they may involve the summation of two or more terms where not all terms involve f_t.

11.3 Special Cases

When dealing with over-dispersion in the manner described in the previous section, there are three special cases that we will briefly describe for scoring the rows and columns of a contingency table. In the multidimensional sense, these scores can then be used as the basis for constructing a visual summary of the association using correspondence analysis. We do not provide all of the features from performing this analysis but instead describe the application of SVD and biplots as a means of visually portraying the scores.

11.3.1 Generalized Poisson Distribution

One way in which over-dispersion can be accommodated for in a two-way contingency table is to assume that the cell joint frequencies (i.e., the values of n_{ij}) come from a generalized Poisson distribution. With γ_1 defined as the dispersion parameter of the distribution, the generalized Pearson residual takes the form

$$\dot{Z}_{ij}^{GP}(\gamma_1|\,f_t) = \frac{n_{ij} - \mathrm{E}\left(n_{ij}|\,f_t;\,\gamma_1\right)}{\sqrt{\mathrm{Var}\left(n_{ij}|\,f_t;\,\gamma_1\right)}}$$

where

$$\mathrm{E}\left(n_{ij} | f_t;\ \gamma_1\right) = \frac{f_t p_{i\bullet} p_{\bullet j}}{1 - \gamma_1} \tag{11.5}$$

and

$$\mathrm{Var}\left(n_{ij} | f_t;\ \gamma_1\right) = \frac{f_t p_{i\bullet} p_{\bullet j}}{\left(1 - \gamma_1\right)^3} \tag{11.6}$$

so that

$$\mathrm{E}\left(p_{ij};\ \gamma_1\right) = \frac{p_{i\bullet} p_{\bullet j}}{1 - \gamma_1}$$

and

$$\mathrm{Var}\left(p_{ij};\ \gamma_1\right) = \frac{p_{i\bullet} p_{\bullet j}}{\left(1 - \gamma_1\right)^3}$$

which are both independent of the sample size, f_t. By adjusting the mean and variance of n_{ij} from $f_t p_{i\bullet} p_{\bullet j}$ to Eqs. (11.5) and (11.6), respectively, the generalized Poisson residual can be expressed as

$$Z_{ij}^{GP}\left(\gamma_1\right) = \frac{p_{ij} - p_{i\bullet} p_{\bullet j} / \left(1 - \gamma_1\right)}{\sqrt{p_{i\bullet} p_{\bullet j} / \left(1 - \gamma_1\right)^3}}.$$

Therefore, a correspondence analysis of the two-way contingency table can be performed by applying a singular value decomposition to

$$Z_{ij}^{GP}\left(\gamma_1\right) = \sum_{k=1}^{K} \tilde{x}_{ik} \rho_k \tilde{y}_{jk} \tag{11.7}$$

where $\tilde{x}_{ik} = \tilde{x}_{ik}\left(\gamma_1\right)$ is the ith element of the kth left (row) singular vector score, $\tilde{y}_{jk} = \tilde{y}_{jk}\left(\gamma_1\right)$ is the jth element of the kth right (column) singular vector score, $\rho_k = \rho_k\left(\gamma_1\right)$ is a singular value and $K = \min(m,\ n) - 1$.

For more information on this residual, one may refer to, for example, Consul (1989), Consul and Jain (1973) and Tripathi and Gupta (1984). When the dispersion parameter is $\gamma_1 = 0$ then $\mathrm{E}\left(p_{ij};\ \gamma_1 = 0\right) = \mathrm{Var}\left(p_{ij};\ \gamma_1 = 0\right)$ so that there is no dispersion present in the data. Conversely, $\gamma_1 > 0$ and $\gamma_1 < 0$ reflect over-dispersion and under-dispersion, respectively. Since the purpose here is to accommodate for over-dispersion in a two-way contingency table, we shall be examining the impact of this residual for values of γ_1 that exceed zero.

11.3.2 Negative Binomial Distribution

A second approach that can be used to accommodate for the presence of over-dispersion in a two-way contingency table is to treat the n_{ij} values being a random variable from a negative binomial distribution (Fisher 1941). In this case, the expectation and variance of n_{ij} is

$$\mathrm{E}\left(n_{ij} \mid f_t;\ \gamma_2\right) = f_t p_{i\bullet} p_{\bullet j}$$

and

$$\mathrm{Var}\left(n_{ij} \mid f_t;\ \gamma_2\right) = f_t p_{i\bullet} p_{\bullet j} + \frac{\left(f_t p_{i\bullet} p_{\bullet j}\right)^2}{\gamma_2},$$

respectively. Therefore

$$\mathrm{E}\left(p_{ij} \mid f_t;\ \gamma_2\right) = p_{i\bullet} p_{\bullet j}$$

and

$$\mathrm{Var}\left(p_{ij} \mid f_t;\ \gamma_2\right) = p_{i\bullet} p_{\bullet j} + f_t \frac{\left(p_{i\bullet} p_{\bullet j}\right)^2}{\gamma_2}$$

and leads to the Negative Binomial residual

$$Z_{ij}^{NB}(\gamma_2) = \frac{p_{ij} - p_{i\bullet} p_{\bullet j}}{\sqrt{p_{i\bullet} p_{\bullet j} + (f_t/\gamma_2)\left(p_{i\bullet} p_{\bullet j}\right)^2}}. \tag{11.8}$$

Over-dispersion arises for those values of $\gamma_2 > 0$ and as $\gamma_2 \to \infty$ then $\mathrm{Var}\left(p_{ij};\ \gamma_2,\ f_t\right) \to \mathrm{E}\left(p_{ij};\ \gamma_2,\ f_t\right)$. Using this result, correspondence analysis may be performed by applying a SVD to the Negative Binomial residuals

$$Z_{ij}^{NB}(\gamma_2) = \sum_{k=1}^{K} \tilde{x}_{ik} \rho_k \tilde{y}_{jk}$$

where $\tilde{x}_{ik} = \tilde{x}_{ik}(\gamma_2)$, $\tilde{y}_{jk} = \tilde{y}_{jk}(\gamma_2)$ and $\rho_k = \rho_k(\gamma_2)$ and $K = \min(m,\ n) - 1$.

11.3.3 Conway-Maxwell Poisson Distribution

Another variant of the Poisson distribution that accommodates for dispersion is the Conway-Maxwell Poisson (CMP) distribution; see Conway and Maxwell (1962). For such a distribution, there is no closed form for the expectation and variance of n_{ij}, however, they may be approximated (Sellers et al., 2012) by

$$\mathrm{E}\left(n_{ij}|\ f_t;\ \gamma_3\right) \approx \left(f_t p_{i\bullet} p_{\bullet j}\right)^{1/\gamma_3} - \frac{\gamma_3 - 1}{2\gamma_3}$$

$$\mathrm{Var}\left(n_{ij}|\ f_t;\ \gamma_3\right) \approx \frac{\left(f_t p_{i\bullet} p_{\bullet j}\right)^{1/\gamma_3}}{\gamma_3}$$

where $\gamma_3 > 1$ reflects under-dispersion in the table and $0 < \gamma_3 < 1$ reflects over-dispersion. Note that, when $\gamma_3 = 1$, there is zero dispersion since $\mathrm{E}\left(n_{ij};\ \gamma_3 = 1\right) = \mathrm{Var}\left(n_{ij};\ \gamma_3 = 1\right)$. It was shown by Shmueli et al. (2005) that such approximations work well when $\gamma_3 \leq 1$ or when $np_{i\bullet}p_{\bullet j} > 10^{\gamma_3}$. In this case, the Conway-Maxwell Poisson residual is

$$Z_{ij}^{CM}(\gamma_3) = \frac{p_{ij} - f_t^{(1-\gamma_3)/\gamma_3}\left(p_{i\bullet}p_{\bullet j}\right)^{1/\gamma_3} + (\gamma_3 - 1)/(2f_t\gamma_3)}{\sqrt{f_t^{(1-\gamma_3)/\gamma_3}\left(p_{i\bullet}p_{\bullet j}\right)^{1/\gamma_3}/\gamma_3}} \tag{11.9}$$

and correspondence analysis of over-dispersed cross-classified data may be performed by considering the SVD of Z_{ij}^{CM} such as

$$Z_{ij}^{CM}(\gamma_3) = \sum_{k=1}^{K} \tilde{x}_{ik}\rho_k\tilde{x}_{jk}\ . \tag{11.10}$$

where $\tilde{x}_{ik} = \tilde{x}_{ik}(\gamma_3)$, $\tilde{y}_{jk} = y_{jk}(\gamma_3)$ and $\rho_k = \rho_k(\gamma_3)$ and $K = \min(m,\ n) - 1$.

11.4 Over-Dispersion, the Biplot and a Re-Examination of Table 3.5

In this section, we explore the correspondence analysis of a two-way contingency table using the residuals described in the previous section. Following on from our discussions in Chaps. 9 and 10, we shall do so by visually summarizing the association between the categorical variables of the table using the row isometric biplot.

For our three residuals, we shall consider different values of the over-dispersion parameters γ_1, γ_2 and γ_3. Indeed, the analyst should be aware that a lot of attention needs to be made on the choice of over-dispersion parameters because, as we shall describe, they do impact on the numerical and visual summaries, and therefore, on the interpretation of the resulting biplots; see for an introduction to biplots Chap. 9.

Suppose we consider the two-way contingency table of Table 3.5 which is reproduced here as Table 11.1. It is formed from the cross-classification of *Perception* of the 11 Rorschach inkblots and six *Mood* levels they induce for a sample of 342 subjects; see also Garmize and Rychlak (1964) and Nishisato (1984, 1985, 1987, 1991) who studied this contingency table. The 11 nominal categories of the *Perception* variable are *Bat*, *Blood*, *Butterfly*, *Cave*, *Clouds*, *Fire*, *Fur*, *Mask* , *Mountains*, *Rocks*

Table 11.1 Subjects' perception of the Rorschach inkblots and their induced mood

Rorschach	Mood					
Perception	Fear	Anger	Depression	Love	Ambition	Security
Bat	33	10	18	1	2	6
Blood	10	5	2	1	0	0
Butterfly	0	2	1	26	5	18
Cave	7	0	13	1	4	2
Clouds	2	9	30	4	1	6
Fire	5	9	1	2	1	1
Fur	0	3	4	5	5	21
Mask	3	2	6	2	2	3
Mountains	2	1	4	1	18	2
Rocks	0	4	2	1	2	2
Smoke	1	6	1	0	1	0

and *Smoke* while the six *Mood* levels of the subjects are *Fear*, *Anger*, *Depression*, *Love*, *Ambition* and *Security*.

Before visualizing the association between the two variables of Table 11.1, we first perform a chi-squared test of independence. Doing so yields a Pearson chi-squared statistic of 370.85 and, with 50 degrees of freedom, has a p-value that is less than 0.001. Therefore, there is strong evidence of a statistically significant association between the two categorical variables.

To help us detect whether Table 11.1 exhibits over-dispersion, we shall be adopting the strategy of Efron (1992)—and briefly described in Sect. 11.1—and construct what we refer to as an *Efron plot*; this plot is given by Fig. 11.1. It shows that there is more variation in the residuals, Z_{ij} for large values of $\sqrt{\mathrm{E}\left(n_{ij}\right)} = \sqrt{f_t p_{i\bullet} p_{\bullet j}}$ than there is for smaller values of $\sqrt{\mathrm{E}\left(n_{ij}\right)}$ suggesting the possibility of over-dispersion being observed in the contingency table. It is not a very strong sign of over-dispersion but enough for us to demonstrate the utility of the strategies outlined earlier in Sect. 11.3.

To address the presence of over-dispersion in Table 11.1, we consider the generalized Poisson residuals for $0 < \gamma_1 < 1$. We also consider values of $\gamma_2 > 0$ for the negative binomial residuals and $0 < \gamma_3 < 1$ for the residuals derived from the Conway-Maxwell Poisson distribution.

In the following sections, we shall be constructing a row isometric biplot using some of the residuals we have described above. For their construction, we apply the SVD to each of the residuals yielding the matrix of row scores, $\tilde{\mathbf{X}}$ (where the kth column is $\tilde{\mathbf{x}}_k$ so that $\tilde{\mathbf{X}}^T\tilde{\mathbf{X}} = \mathbf{I}$), the matrix of column scores, $\tilde{\mathbf{Y}}$ (where the kth column is $\tilde{\mathbf{y}}_k$ so that $\tilde{\mathbf{Y}}^T\tilde{\mathbf{Y}} = \mathbf{I}$) and the matrix of singular values. Squaring these singular values gives the principal *inertia values* for each of the $K = \min(11, 6) - 1 = 5$ dimensions of the biplot and are summarized in the following tables. Table 11.2 summarizes these values for various values of γ_1 for the generalized Poisson residual,

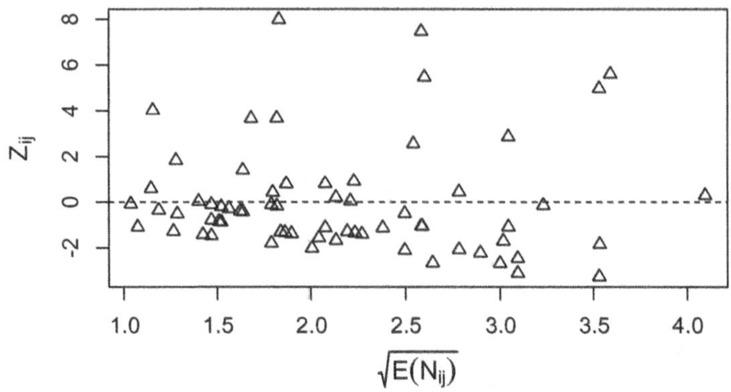

Fig. 11.1 Efron's plot of Table 11.1

Table 11.2 The five principal inertia values of Table 11.1 using various γ_1 values for the generalized Poisson residuals

Inertia values	$\gamma_1 = 0.1$	$\gamma_1 = 0.25$	$\gamma_1 = 0.55$	$\gamma_1 = 0.75$	$\gamma_1 = 0.9$
ρ_1^2	0.338	0.195	0.136	0.141	0.081
ρ_2^2	0.183	0.106	0.042	0.007	0.000
ρ_3^2	0.124	0.072	0.023	0.004	0.000
ρ_4^2	0.093	0.054	0.016	0.003	0.000
ρ_5^2	0.053	0.047	0.012	0.002	0.000
Total Inertia	0.791	0.474	0.228	0.156	0.082
X^2	270.353	162.078	78.096	53.503	28.048
p-value	<0.001	<0.001	0.007	0.341	0.995

while Table 11.3 summarizes those values for various γ_2 values that define the negative binomial distribution. Finally, Table 11.4 summarizes the principal inertia values for the five dimensions of the isometric biplot when using various values of γ_3 to define the Comway-Maxwell residuals.

Table 11.2 show as γ_1 increases, the total inertia of the contingency table decreases as does the chi-squared statistic, $\tilde{X}^2(\gamma_1)$. In fact, for the generalized Poisson residual, as γ_1 approaches zero, $\tilde{X}^2(\gamma_1)$ approaches the Pearson statistic for Table 11.1. The value of total inertia decreases to zero as γ_1 gets closer to 1. In fact, the p-value of $\tilde{X}^2(\gamma_1)$ is close to 0.01 for values of γ_1 that are close to 0.55. When considering the negative binomial residuals, Table 11.3 shows that as γ_2 increases, so does the principal inertia along each dimension of the biplot that can be constructed. Therefore, as γ_2 increases, the resulting value of $\tilde{X}^2(\gamma_2)$ gets closer to the Pearson chi-squared statistic of independence; for all values of γ_2, we have considered in Table 11.3, the p-value of the $\tilde{X}^2(\gamma_2)$ values are all very small. These findings should not be surprising, given the statements we have made above for the negative binomial residual. Finally, Table 11.4, summarizes the principal inertia values, total inertia, $\tilde{X}^2(\gamma_3)$ values, and

Table 11.3 The five principal inertia values of Table 11.1 using various γ_2 values for the negative binomial residuals

Inertia values	$\gamma_2 = 5$	$\gamma_2 = 10$	$\gamma_2 = 15$	$\gamma_2 = 20$	$\gamma_2 = 25$
ρ_1^2	0.196	0.268	0.309	0.336	0.355
ρ_2^2	0.140	0.180	0.199	0.209	0.217
ρ_3^2	0.087	0.109	0.121	0.128	0.134
ρ_4^2	0.057	0.080	0.092	0.100	0.105
ρ_5^2	0.031	0.044	0.050	0.055	0.057
Total Inertia	0.512	0.682	0.772	0.828	0.867
$\tilde{X}^2\ (\gamma_2)$	175.104	233.244	264.024	283.176	296.514
p-value	<0.001	<0.001	<0.001	<0.001	<0.001

Table 11.4 The five principal inertia values of Table 11.1 using various γ_3 values for the Conway-Maxwell Poisson residuals

Inertia values	$\gamma_3 = 0.25$	$\gamma_3 = 0.55$	$\gamma_3 = 0.75$	$\gamma_3 = 0.90$	$\gamma_3 = 0.95$
ρ_1^2	195.535	1.884	0.349	0.336	0.397
ρ_2^2	0.007	0.063	0.183	0.195	0.222
ρ_3^2	0.002	0.047	0.121	0.126	0.147
ρ_4^2	0.001	0.035	0.079	0.101	0.114
ρ_5^2	0.000	0.016	0.056	0.053	0.062
Total Inertia	198.544	2.046	0.788	0.810	0.942
$\tilde{X}^2\ (\gamma_3)$	67902.18	699.621	269.587	276.855	322.036
p-value	<0.001	<0.001	<0.001	<0.001	<0.001

its p-value for five values of γ_3 used to define the Conway-Maxwell residuals. This table shows that as γ_3 increases, the behaviour of the principal inertia values is inconsistent. For example, the value of ρ_1^2 decreases, but then increases for γ_3 values exceeding 0.8. However, for ρ_2^2, it decreases immediately from near infinity for very small values of γ_3, but then increase from $\gamma_3 = 0.55$ onwards. The impact this has on the total inertia and the value of $\tilde{X}^2\ (\gamma_3)$ is that they both decrease as γ_3 moves from zero to about 0.9, then increases so that as γ_3 approaches 1, and the value of $\tilde{X}^2\ (\gamma_3)$ approaches the Pearson chi-squared statistic of Table 11.1.

Since we have been able to calculate the principal inertia values of Table 11.1, we can also determine the percentage contribution they make to the total inertia of the contingency table. To show this, Table 11.5 summarizes these percentage contributions for $\gamma_1 = 0.55$ (when using the generalized Poisson residuals), $\gamma_2 = 15$ (for the negative binomial residuals) and $\gamma_3 = 0.55$ (for the Conway-Maxewell residuals). As expected, the first principal inertia accounts for most of the association that exists between the variables of the contingency table, but they do not account for the same amount of the association.

Table 11.5 Percentage of the total inertia accounted of Table 11.1 for by each of the five principal inertia values using the classic Pearson residuals, the generalized Poisson residual ($\gamma_1 = 0.55$), the negative binomial residual ($\gamma_2 = 15$), and the Conway-Maxwell Poisson residuals ($\gamma_3 = 0.55$)

Perc. Inertia	Classic	Generalized Poisson ($\gamma_1 = 0.55$)	Negative Binomial ($\gamma_2 = 15$)	Conway-Maxwell Poisson ($\gamma_3 = 0.55$)
ρ_1^2	42.724	59.609	40.093	92.121
ρ_2^2	23.098	18.487	25.747	3.081
ρ_3^2	15.711	9.995	15.643	2.279
ρ_4^2	11.814	6.798	11.974	1.716
ρ_5^2	6.652	5.112	6.542	0.804
2D Biplot	65.822	78.096	65.840	95.202

Table 11.5 shows that when assessing the impact of the over-dispersion parameter on the quality of a joint depiction of the row and column categories of Table 11.1, different residuals lead to different results. Of course, while other values of γ_1, γ_2 and γ_3 are not reflected in these percentages, different values of these parameters also impact upon the quality of the display. It shows that for $\gamma_1 = 0.55$, the principal inertia values along each dimension of a joint display are higher than those of the principal inertia values when performing the classical approach to correspondence analysis. Interestingly, Table 11.5 also shows that the percentage contribution of each principal inertia to the total inertia for γ_2, when using the negative Binomial distribution, is quite similar to the percentages when using the classical (standardized) residual or, equivalently, the generalized Pearson residual. Where differences become noticeable in making these comparisons is when, for the Conway-Maxwell Poisson residuals, a value of $\gamma_3 = 0.55$ produces very different percentage contribution values. In fact, it shows that when confining a joint display to two dimensions, applying a SVD to the matrix of Conway-Maxwell residuals when $\gamma_3 = 0.55$ produces a display that accounts for about 95% of the association between the two variables of Table 11.1; where the first dimension accounts for a massive 92.1% of the total inertia and the second dimension accounts for only 3.1%. Such a joint display is far superior in quality when compared with the two-dimensional displays obtained when using the classic (standardized) residual, the generalized Poisson residual (with $\gamma_1 = 0.55$) or the negative binomial residual (with $\gamma_2 = 15$).

Now that we have provided an overview of the quality of a two-dimensional display that comes from applying a SVD to each of the residuals, we now turn our attention to investigate the configurations of points that are produced. For example, Fig. 11.2 gives the classic row isometric biplot of Table 11.1; this is obtained by performing a SVD on the matrix of Pearson residuals that is defined by Eq. (11.1). It shows that the Rorschach inkblot that is perceived to be the *Butterfly* is strongly associated with *Security* and *Love*. It also shows that inkblots perceived to represent a *Mountain* are strongly linked to *Ambition*. One can also see that inkblots perceived to look like *Blood*, *Bats*, and *Fire* are associated with *Fear* and *Anger*. A similar

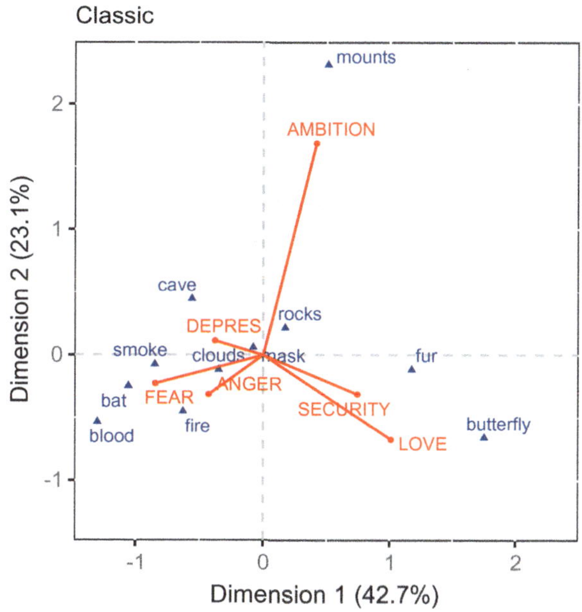

Fig. 11.2 Row isometric biplot Table 11.1 via the SVD of the Pearson residual, Eq. (11.1)

configuration can also be obtained when applying a SVD to the matrix of generalized Poisson residuals when $\gamma_1 = 0.55$.

Figure 11.3 gives the two-dimensional biplot of Table 11.1, when a correspondence analysis is performed by applying a SVD to the matrix of negative binomial residuals defined by Eq. 11.8 when $\gamma_2 = 15$. Other than some very minor differences in the configuration of points, Fig. 11.3 gives a display that is virtually the same as the configuration of points given in Fig. 11.2. Therefore, the association between the two variables is similarly depicted in two dimensions. As Table 11.5 suggests, the quality of the plot is very similar, with Fig. 11.3 displaying only half a percent more of the total inertia than Fig. 11.2.

Interestingly, while Table 11.5 shows that a two-dimensional row isometric biplot of Table 11.1 obtained by applying a SVD to the matrix of Conway-Maxwell Poisson residuals accounts for nearly 95% of the association, the configuration of points are NOT centred at the origin of the display. Therefore, the biplot is not included here. The cause of this unwanted feature is uncertain. Our experience with working with the residual for the purposes of performing a correspondence analysis suggest that the cause may be due to the presence of a trivial solution that is "randomly" placed in a column of $\tilde{\mathbf{X}}$ and/or $\tilde{\mathbf{Y}}$. We use the term "randomly" here very loosely although different values of γ_3 will produce a trivial solution in a different column to either/both of these matrices.

The presence of this phenomenon shows that it is important to understand how and why a joint display is constructed for visualizing the association between the categorical variables of a contingency table. Nishisato has argued throughout his

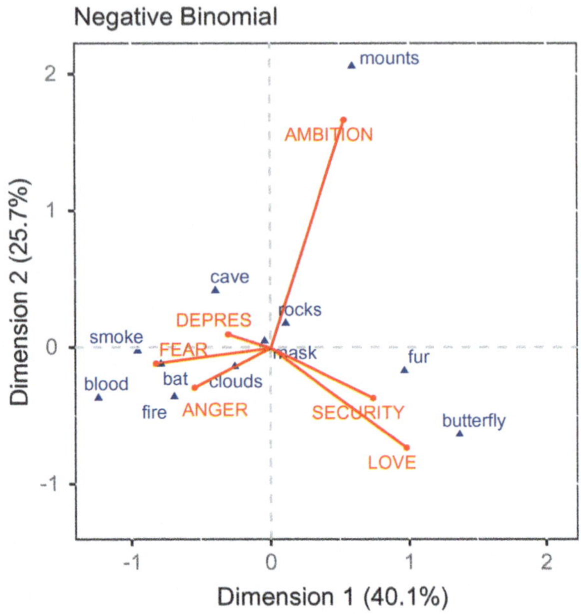

Fig. 11.3 Row isometric biplot Table 11.1 via the SVD of the negative binomial residual, Eq. (11.8), with $\gamma_2 = 15$

career, and most recently with Clavel, that such caution is indeed warranted and has detailed situations, where such a joint display is reasonable and when a joint display produces spurious conclusions about the nature of the association. Their arguments in Part I of this book continue this discussion.

While Beh and Lombardo (2014, 2020b) and many others in the world of statistics, and her allied disciplines, have extensively used and endorsed the joint visualization of association between categorical variables, this is undertaken by ensuring specific features of such displays are preserved. For example, irrespective of the method used, it is important that such displays are meaningful and interpretable. From a correspondence analysis perspective, it is highly desirable that the proximity of row (and column) coordinates preserve the *property of distributional equivalence*, that the origin has a clear and meaningful interpretation and that two-dimensional (say) display depicts a reasonable amount of the association that exists between the categorical variables; in the past we have advocated that such a display depict at least 70% of this association to be of any use.

These comments suggest that dealing with over-dispersion in a contingency table in the manner we have so far described is not without its faults (like many statistical techniques). So, before ending our discussions in this book, we provide an alternative strategy for dealing with over-dispersion. Rather than doing so, by introducing a dispersion parameter like we have done above, one may instead stabilize the variance in some way. Here we describe the *adjusted standardized residual* and the *Freeman-Tukey residual*.

11.5 Stabilizing the Variance

11.5.1 The Adjusted Standardized Residual

One further strategy that may be adopted to accommodate for the presence of over-dispersion in a contingency table is to consider the adjusted Pearson residuals proposed by Haberman (1973). This residual for the (i, j)th cell of the table is

$$Z_{ij}^A = \frac{Z_{ij}^*}{f_t} = \frac{p_{ij} - p_{i\bullet}p_{\bullet j}}{\sqrt{p_{i\bullet}p_{\bullet j}\left(1 - p_{i\bullet}\right)\left(1 - p_{\bullet j}\right)}} \tag{11.11}$$

and takes into account that the variance of $\sqrt{f_t}Z_{ij}$ is $\left(1 - p_{i\bullet}\right)\left(1 - p_{\bullet j}\right)$; see Haberman (1973). For further details, the interested reader can refer to Beh (2012), who described some of the technical issues and gave a practical demonstration of using this residual for performing a correspondence analysis on a two-way contingency table. By following a similar strategy described above for the various residuals presented in this chapter, over-dispersion in a contingency table can be addressed by applying a singular value decomposition to the adjusted standardized residual given by Eq. (11.11).

11.5.2 The Freeman-Tukey Residual

Since over-dispersion arises because the variance of n_{ij} is larger than its expectation, a popular method used to stabilize its variance is to consider the transformation proposed by Freeman and Tukey (1950), who, adapting their notation for a two-way contingency table, showed that

$$\sqrt{n_{ij}} + \sqrt{n_{ij} + 1}.$$

has a mean and variance of $\sqrt{4np_{i\bullet}p_{\bullet j} + 1}$ and 1, respectively. Thus, Freeman and Tukey (1950) proposed

$$\tilde{T}^2 = \sum_{i=1}^{I}\sum_{j=1}^{J}\left(\sqrt{n_{ij}} + \sqrt{n_{ij} + 1} - \sqrt{4np_{i\bullet}p_{\bullet j} + 1}\right)^2$$

as an alternative to Pearson's chi-squared statistic. Note that $\tilde{T}^2$ may be alternatively, but equivalently, expressed as

$$\tilde{T}^2 = n\sum_{i=1}^{I}\sum_{j=1}^{J}\left(\sqrt{p_{ij}} + \sqrt{p_{ij} + \frac{1}{n}} - \sqrt{4p_{i\bullet}p_{\bullet j} + \frac{1}{n}}\right)^2 .$$

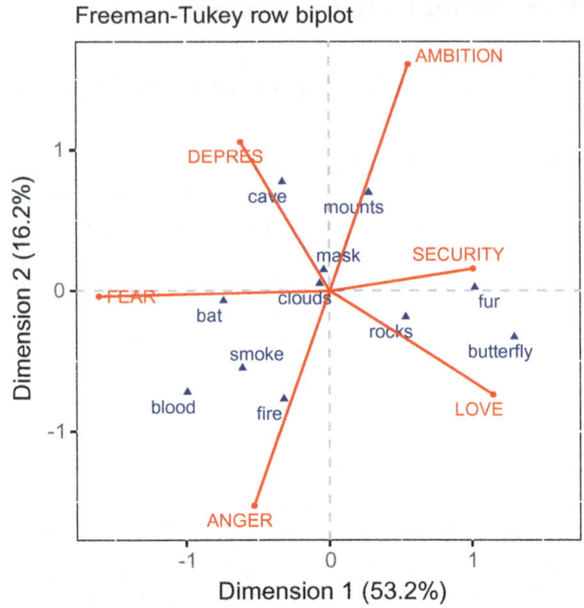

Fig. 11.4 Row isometric biplot Table 11.1 via the SVD of the Freeman-Tukey residual, Eq. (11.12)

For large sample sizes, $\tilde{T}^2$ may be approximated by eliminating $1/n$. Doing so gives the *Freeman-Tukey* statistic

$$\breve{T}^2 = 4n \sum_{i=1}^{I} \sum_{j=1}^{J} \left(\sqrt{p_{ij}} - \sqrt{p_{i\bullet}p_{\bullet j}}\right)^2 .$$

Therefore, scoring the rows and column can be made via a singular value decomposition

$$2\left(\sqrt{p_{ij}} - \sqrt{p_{i\bullet}p_{\bullet j}}\right) = \sum_{k=1}^{K} \tilde{x}_{ik}\rho_k\tilde{y}_{jk} . \tag{11.12}$$

This approach for dealing with over-dispersion in correspondence analysis was described by Beh et al. (2018).

Figure 11.4 shows the row isometric biplot Table 11.1, where a SVD is applied to the matrix of Freeman-Tukey residuals, defined by Eq. (11.12). While the classic row isometric biplot obtained by performing a SVD on the generalized Pearson residuals, visually describes about 65% of the association between the variables, the labels of Fig. 11.4 shows that this two-dimensional biplot depicts about 69% of the association; so, for this table, performing a correspondence analysis by decomposing the Freeman-Tukey residual produces a better quality display than applying a SVD to the matrix of Z_{ij} values. As for the configuration of points, the interpretation is similar to that of Fig. 11.2, although Fig. 11.4 shows that the *Mood* categories appear more evenly distributed across the two dimensions. It also suggests that inkblots perceived to be *Smoke* are equally associated with *Fear* and *Anger*, while Fig. 11.2 suggests that *Smoke* is more strongly associated with *Fear*, followed by *Depression*. Figure 11.2 suggests that, while *Rocks* lies close to the origin, it is associated with *Ambition*, while Fig. 11.4 indicates that this inkblot pattern is more strongly liked to *Love*.

11.6 Final Remarks

In this chapter, we have shown that when scoring or, from a visual perspective, performing correspondence analysis is undertaken and over-dispersion arises in a contingency table (i.e., when $\text{Var}\left(n_{ij}\right) > \text{E}\left(n_{ij}\right)$) strategies can be put in place to accommodate for this additional variation that classical techniques do not consider. One such strategy has been to incorporate a dispersion parameter into different residual measures. This chapter showed how the generalized Poisson residual, negative binomial residual, and Conway-Maxwell Poisson residual may be used in this situation. The advantage of considering these three residuals is that when there is zero dispersion then they simplify to the classic standarized residual, Z_{ij}. For some of these residuals, the resulting biplot produces an interpretable visual depiction of the association, although we concede that further work needs to be undertaken to understand some of their features. This highlights to us the need to understand how to construct and interpret visual depictions of association, but also to know that care must be taken not to be too hasty in assuming such visual summaries are "perfect".

Rather than introducing a dispersion parameter, a second strategy for dealing with over-dispersion is simply to stabilize the variance. We have briefly outlined that two appropriate residuals to consider in doing so are the adjusted standardized residual of Haberman (1973) and the Freeman-Tukey residual of Freeman and Tukey (1950). From a correspondence analysis perspective, these residuals have been used in the development and application of the work described in Beh (2012) and Beh et al. (2018). They are relatively simple approaches and provide an interpretable visual depiction of the association without any computational or technical hurdles to overcome.

Despite the benefits, or otherwise, of these strategies, further work can be undertaken to improve the applicability and development of these tools for scoring or visualizing the association between categorical variables. In particular, they have yet to be extended to the case where one is interested in accommodating for the presence of over-dispersion in a three-way, or multi-way, contingency table. Nor have they been described when there exists an asymmetric association between the categorical variables.

This book has provided an overview of a range of issues concerned with the scoring and visualization of categorical variables. As we have said from the outset of the book, the intention was not to necessarily provide anything new to the reader, but to balance the argument that has long involved discussing the pro's and con's of visualizing the association between categorical variables. While the focus of this book has been predominately on the perspectives of the four of us as authors, and with particular emphasis on the long and enduring career of Nishisato, there is an understanding that with the advance of time the next generation of researchers will always question, critique, and advance the knowledge of those that came before them in all areas of sciences; the discipline of statistics is no exception to this.

References

Agresti, A. (2013). *Categorical Data Analysis* (3rd ed.). New York: Wiley.

Beh, E. J. (2012). Simple correspondence analysis using adjusted residuals. *Journal of Statistical Planning and Inference, 142,* 965–973.

Beh, E. J., & Lombardo, R. (2020a). Five strategies for accommodating overdispersion in simple correspondence analysis. In Imaizumi, T., Okada, A., Miyamoto, S., Sakaori, F., Y. Yamamoto, Y., & Vichi, M., (Eds.), *Advanced Studies in Classification and Data Science. Studies in Classification, Data Analysis, and Knowledge Organization* (pp. 117–129). Singapore: Springer.

Beh, E. J., Lombardo, R., & Alberti, G. (2018). Correspondence analysis and the freeman-tukey statistic: A study of archaeological data. *Computational Statistics and Data Analysis, 128,* 73–86.

Beh, E. J., & Lombardo, R. (2014). *Correspondence analysis: Theory practice and new strategies.* Chichester: Wiley.

Beh, E. J., & Lombardo, R. (2020b). *An introduction to correspondence analysis.* Chichester (in press): Wiley.

Consul, P. (1989). *Generalized poisson distributions: Properties and applications.* New York: Marcel Dekker.

Consul, P. C., & Jain, G. C. (1973). A generalization of the Poisson distribution. *Technometrics, 15,* 791–799.

Conway, R. W., & Maxwell, W. (1962). A queuing model with state dependent service rates. *Journal of Industrial Engineering, 12,* 132–136.

Efron, B. (1992). Over dispersion estimates based on the method of asymmetric maximum likelihood. *Journal of the American Statistical Association, 87,* 98–107.

Fisher, R. A. (1941). The negative binomial distribution. *Annals of Eugenics, 11,* 182–187.

Freeman, M. F., & Tukey, J. W. (1950). Transformations related to the angular and square root. *The Annals of Mathematical Statistics, 21,* 607–611.

Garmize, L. M., & Rychlak, J. F. (1964). Role-play validation of a sociocultural theory of symbolism. *Journal of Consulting Psychology, 28,* 107–115.

Haberman, S. (1973). The analysis of residuals in cross-classified tables. *Biometrics, 75,* 457–467.

Nishisato, S. (1984). Forced classification: A simple application of a quantification method. *Psychometrika, 49*(1), 25–36.

Nishisato, S. (1985). Methods for handling outlier responses in dual scaling. In *Proceedings of the 49th Annual Meeting of the Japanese Psychological Association,* (p. 50)?

Nishisato, S. (1987). Robust techniques for quantifying categorical data. In MacNeil, B. & Umphrey, G. J., (Eds.), *Foundations of statistical inference.* (pp. 209–217). Dordrecht: D. Reidel Publishing.

Nishisato, S. (1991). Standardizing multidimensional space for dual scaling. In *Proceedings of the 20th Annual Meeting of the German Operations Research Society, Hohenheim University,* (pp. 584–591).

Sellers, K. F., Borle, S., & Schmueli, G. (2012). The COM-Poisson model for count data: A survey of methods and applications. *Applied Stochastic Models in Business and Industry, 28,* 104–116.

Shmueli, G., Minka, T. P., Kadane, J. B., Borle, S., & Boatwright, P. (2005). A useful distribution for fitting discrete data: Revival of the conway-maxwell-poisson distribution. *Applied Statistics, 54,* 127–142.

Tripathi, R. C., & Gupta, R. C. (1984). Statistical inference regarding the generalized Poisson distribution. *Sankhya (Series B), 46,* 166–173.

The manufacturer's authorised representative in the EU is Springer Nature Customer Service Centre GmbH, Europaplatz 3, 69115 Heidelberg, Germany. If you have any concerns regarding our products, please contact ProductSafety@springernature.com

Printed and bound by CPI Group (UK) Ltd, Croydon, CR0 4YY
30/09/2026
02225330-0007